Atlantic Wildflowers

Atlantic Wildflowers

Photography by
WAYNE BARRETT & ANNE MacKAY

Text by
DIANE GRIFFIN

Toronto
OXFORD UNIVERSITY PRESS
1984

ACKNOWLEDGEMENTS

I wish to thank a number of people and organizations who assisted in the preparation of this book. Sandra Englehardt of the Alberta Energy and Natural Resources Library provided able assistance with references for many species. Kevin Griffin helped prepare the manuscript. Dr. S.P. Vander Kloet of Acadia University provided support, including the use of the E.C. Smith Herbarium. Dr. Ian MacQuarrie, Dr. Lawson Drake, and Daryl Guignion of the University of Prince Edward Island gave encouragement, as well as the use of university facilities.
The following people typed the manuscript: Candas Jane Dorsey, Donna Knutson, Ivy Robinson, Sheri Priest, and Alicia Sia.
Thanks also to Sharleen Griffin and Errol and Margaret Laughlin for their support and encouragement, and to Roger Boulton for his editorial assistance.

D.G.

Canadian Cataloguing in Publication Data
Griffin, Diane.
Wildflowers of the Atlantic Provinces

Includes index.
ISBN 0-19-540463-7

1. Wild Flowers – Atlantic Provinces – Identification.
I. Barrett, Wayne. II. MacKay, Anne. III. Title.

QK203.A7B37 1984   582.12′09715   C84-098655-6

Produced by Roger Boulton Publishing Services,
Designed by Fortunato Aglialoro

© 1984 Oxford University Press (Canadian Branch)

1 2 3 4 – 7 6 5 4
Printed in Hong Kong by Scanner Art Services, Inc.,
Toronto

INTRODUCTION

When the Acadians arrived in the three Maritime provinces, most of the land was forested, so they dyked salt marshes and used the native marsh hay for their livestock. As more land became cleared and planted, weed seeds were present among those of the desired crop. Many areas, such as the mountains of Newfoundland, the Cape Breton Plateau, and bogs, were not well explored botanically until later when the emphasis was on agriculture and beating back the wilderness. Nevertheless, the wilderness was still appreciated for its bounty of wild fruit and for other useful portions of plants. With the increased dependence on cultivated species and imported food, the wild plants became less important as a food source. During the past two decades, however, wild plants, both native and introduced, have been rediscovered for food purposes as well as for their natural beauty.

This book examines a representative sample of plants that grow in Atlantic Canada in a variety of habitats, from the seashore to rich deciduous woodlands. Some grow along roadsides or invade gardens, requiring great effort to beat back. Many of these, such as the mustards, were imported as seeds in shipments of grain or in ballast used in the days of wooden sailing ships. Ragwort and other species are still spreading rapidly within the region. A few, such as the lupin, were planted for ornamental purposes and have since escaped to invade new territory.

Atlantic Canada today consists of Nova Scotia, Prince Edward Island, New Brunswick, Newfoundland, and Labrador. Most species found there extend westwards into other provinces and southwards into the United States. Since the Atlantic area was one of the earliest settled parts of Canada, a large number of floral collections have been made and housed in herbaria throughout North America. Most of these species are represented in the E.C. Smith Herbarium at Acadia University.

The text for each plant examined here is organized on the following key pieces of information:

COMMON NAME
Although plants have common names known to many people, they often vary from one region or province to another, so only the most commonly used names are listed here. The sources for most names are *The flora of Nova Scotia* by A.E. Roland and E.C. Smith, and *Gray's manual of botany*, eighth edition, by M.L. Fernald.

LATIN NAME
Every plant has a scientific name ascribed to that plant alone. After acceptance by the International Botanical Congress, this name becomes the one by which

the plant is known world-wide, helping prevent the confusion created by the variability of common names, and providing a consistent reference base. There are two parts to every name: the genus and the species. An example is the pitcher-plant, *Sarracenia purpurea*, the floral emblem of Newfoundland and Labrador.

The scientific names used in this book follow the nomenclature used in *The flora of Nova Scotia* by Roland and Smith.

GENUS

Species that exhibit similar characteristics are grouped into a genus, which, in the case of the plants studied here, usually contains some species that are not present in Atlantic Canada. The information for the total number of species in each genus came from the *New Britton and Brown illustrated flora of the northeastern United States and adjacent Canada* by H.A. Gleason.

FAMILY

While similar species of plants are grouped into a genus, similar genera are grouped into families. These groupings are primarily based on the structure of the flowers. An example is *Brassicaceae*, the large Mustard family, in which all species normally have four sepals, four petals, six stamens, and a superior ovary with two cells. The Atlantic Canada flora of approximately 2,000 species is grouped into about 114 families.

ETYMOLOGY

The Latin binomial name may have its root in the languages of other cultures. Where this has occurred, it is noted in the text. The origins of the names are given for both the genus and species. Some were named for their medicinal uses, others from the plant's appearance or geographical origin, and still others to honour a person. An example is ragwort, *Senecio Jacobea*, in which the genus name is taken from the Latin *senex*, an old man (referring to the hoariness of many species). *Jacobea* means of St. James. The source for etymologies throughout this book was *Gray's manual of botany*, eighth edition, by Fernald.

HABITAT

Plants usually have a preferred habitat. Some, such as spring-beauty, *Claytonia caroliniana*, occur only in rich hardwood forests, while others, such as the Canada thistle, *Cirsium arvense*, are opportunists that can grow almost anywhere. Each plant has its individual requirements for such factors as soil quality, available nutrients, pH, moisture, salt tolerance, and amount of sunlight.

RANGE

The range is the territory in which a plant species occurs. Some, such as the primrose, *Primula laurentiana*, are only found in areas influenced by the Gulf of St. Lawrence and St. Lawrence River, while others, such as butter-and-eggs, are introduced from Eurasia and have spread throughout most of North America. Many environmental factors affect the range of a species. Most of the range information in this book is based on Roland and Smith's *The flora of Nova Scotia*.

LONGEVITY
Plants are either annuals, biennials, or perennials. Annuals live for only one season and must produce enough seeds to ensure next year's crop. Biennials last two years but produce their flowers and fruits in the second season. Perennials last many seasons and often rely on vegetative reproduction to spread new plants from their underground systems. Some perennials, such as hedge-nettle, have become aggressive weeds that are difficult to control. A few species may belong to more than one category as their longevity can be affected by environmental conditions.

FLOWERING
Because of annual weather variations and other factors that influence growth and flowering, the times listed in this book are to be considered approximate. There can also be considerable variation from the southern to northern part of the range. Plants generally flower much earlier in southern Nova Scotia as compared with northern Newfoundland.

FEATURES
For each species the book gives the general identifying characteristics for identification purposes. More detail can be found in taxonomic reference books.

FRUIT
Flowering plants produce many types of fruit ranging from very simple structures in the grasses to complicated structures in the Pea and other advanced families. The definition for each technical term used is given in the glossary at the back of the book.

HEIGHT
Metric measurements are given for the mature height of each plant, but these can vary because of environmental conditions. A species responds to the favourability of soil, sunlight, and other factors. For example, the field-chickweed, *Cerastium arvense*, will only reach a height of 10 cm. in poor soil but can grow to 25 cm. in favourable conditions.

ECOLOGY
Wild animals depend on plants for protective cover and often for food. Many, such as insects, assist plants with pollination, while others assist in seed

dispersal by accidentally carrying seeds to new locations. Some of these interrelationships are noted.

USES
Wild plants often have value for food or other uses by humans, and these are noted in what follows. Many plants regarded as nuisances to be eradicated from our lawns and fields are in fact good sources of nutrition.

HORTICULTURE
An increasing number of people transplant wildflowers to add beauty and variety to their gardens. This causes little problem as long as the natural conditions are duplicated in the home garden so that the plant will survive. However, one should be very careful to choose only species that are not scarce or endangered in their native habitat. Some species, such as members of the Orchid family, do not transplant well and are better left undisturbed. Further information can be obtained from *Wildflower perennials for your garden* by B. Miles or *Wildflower gardening* by J. Crockett and O. Allen.

SIMILAR SPECIES
Sometimes a number of other species resemble the one under discussion. These are listed to assist the reader in determining differentiation of the plants.

REFERENCES
Mentioned throughout are books and articles that are readily available and that will prove useful for additional information on each species.

PHOTOGRAPHY
All the plants in this book were photographed with 35 mm. Pentax cameras using either Kodachrome 64 or Kodachrome 25 film. Lenses used included 15 mm. to 100 mm. macro lenses, and 28 mm., 35 mm., and 50 mm. lenses. Although various light conditions were encountered, most photos were shot early in the day or under subdued light.

SEQUENCE OF PLANTS
Finally a word as to the arrangement of species within the book. This is in scientific or phylogenetic order rather than by flower colour or in alphabetical order by either common or Latin names. The phylogenetic system attempts to classify plants according to their evolutionary sequence and to their relationships as inferred from indirect evidence or proved by genetic experimentation. This method is vastly superior to any that was used by the early botanists but is still being improved upon as a better understanding develops of the structure and function of a plant's reproductive parts.

Edmonton, January 1984   DIANE GRIFFIN

Broad-leaved cat-tail

Latin Name: *Typha latifolia*
Genus: About 10 species of the temperate and tropical regions of eastern and western hemispheres; 2 in Atlantic Canada
Family: Typaceae (Cat-tail)
Etymology: *Typhe*, the old Greek name; broad-leaved
Habitat: Swamps, shallow ponds, estuaries above salt water, wet fields, ditches, edges of rivers and streams, or in floating bog associations
Range: Throughout North America
Longevity: Perennial
Flowering: Summer
Features: Male flowers with yellow pollen occur on a spike above the female flowers, which form a 2–5 cm. spike in fruit.
Fruit: Nutlets

Height: 1–2.7 m.
Uses: Young shoots can be used in a salad or pickled. The pollen is useful for flour and the roots can be cooked as a vegetable. A very popular emergency food plant.
Similar Species: Forma *ambigua* has the male and female flowers separated by a small distance on the spike. *T. angustifolia* has narrow, slightly rounded leaves.

References: Grace, J.B. and R.G. Wetzel, 1982. Niche differentiation between two rhizomatous plant species: *Typha latifolia* and *Typha angustifolia* (Ecology, morphology U.S.A.). *Canadian Journal of Botany*, vol. 60 (1), 46–57.

Murkin, H.R., R.M. Kaminski, and R.D. Titman, 1982. Responses by dabbing ducks and aquatic invertebrates to an experimentally manipulated cattail marsh (*Typha latifolia*, artificially created cover: water ratios, Canada). *Canadian Journal of Zoology*, vol. 60 (10), 2324–32.

Arrow-grass

Latin Name: *Trilochin elata*
Genus: About 12 species, mostly in the Mediterranean region and Australia; 2 in Atlantic Canada
Family: Juncaginaceae (Arrow-grass)
Etymology: Greek *treis*, three, and *glochis*, point; tall
Habitat: Salt marshes and occasionally in acidic peat around lakes
Range: Labrador to Alaska, south to New Jersey and Mexico; Eurasia
Longevity: Perennial
Flowering: June–July
Features: Numerous small, greenish, bractless flowers grow in a long linear raceme at the top of a naked stout scape. Each has 3 petals and 3 similar sepals. The thick linear leaves are rounded in cross-section and arise from the plant's base.
Fruit: Follicle
Height: Up to 65 cm.
Ecology: Food for waterfowl
Note: The plant is reported to be poisonous for cattle and sheep.
Similar Species: Sometimes this plant is classified as *T. maritima*. *T. palustris* is a smaller plant with linear or club-shaped fruit.

References: Thomas, V.G. and J.P. Prevett, 1980. The nutritional value of arrow-grasses to geese at James Bay. *Journal of Wildlife Management*, vol. 44 (4), 830–6.

Foxtail barley
also called squirrel-tail grass

Latin Name: Hordeum jubatum
Genus: About 20 species of temperate climates; 3 in Atlantic Canada
Family: Poaceae (Grass)
Etymology: The ancient Latin name; with a mane
Habitat: Along the upper limits of salt marshes, along roadsides, farmyards, and dikelands
Range: Labrador to Alaska, south to Virginia and Mexico
Longevity: Annual or biennial
Flowering: June–August
Features: A weedy grass, easily recognized by its squirrel-tail type of spike with long green or purplish awns. The narrow leaves clasp the stem at the joints.
Fruit: Small grain
Height: 30–70 cm.
Uses: The plant is edible before flowering, but in hay the long prickly awns cause serious inflammation in mouths of cattle and horses. It is popular in dried-flower arrangements.
Note: From this plant agronomists have evolved the plump nutritious grains of our cultivated barley.

References: Babbel, G.R. and R.P. Wain, 1977. Genetic structure of (the tetraploid weed) *Hordeum jubatum*. I. Outcrossing rates and heterozygosity levels. *Canadian Journal of Genetic Cytology*, vol. 19 (1), 143–52.

Banting, J.D., 1979. Germination, emergence and persistence of foxtail barley (*Hordeum jubatum*, hazard to livestock). *Canadian Journal of Plant Science*, vol. 59 (1), 35–41.

Murry, L.E. and W. Tai, 1980. Genome relations of *Agropyron sericeum*, *Hordeum jubatum* and their hybrids. *American Journal of Botany*, vol 67 (9), 1374–79.

Ungar, I.A., 1974. The effect of salinity and temperature on seed germination and growth of *Hordeum jubatum*. *Canadian Journal of Botany*, vol. 52 (6), 1357–62.

Marram grass
also called American beach grass or sand reed

Latin Name: *Ammophila breviligulata*
Genus: 3 species, one in Atlantic Canada, the other 2 in Europe
Family: Poaceae (Grass)
Etymology: Greek *ammos*, sand, and *philein*, to love; with short ligule
Habitat: Upper edges of sandy beaches or covering coastal sand dunes
Range: Newfoundland and southern Labrador to North Carolina; and around the Great Lakes; Europe
Longevity: Perennial
Flowering: July–September
Features: Panicle is densely flowered and spike-like. Each spikelet has only one fertile floret, which readily separates and falls away from the glumes. The plant has creeping rhizomes, rigid culms, and involute leaves.

Fruit: Caryopsis
Height: 5–100 cm.
Ecology: Provides fodder for the Sable Island ponies
Uses: Young shoots can be nibbled. The wiry root-stocks would serve in an emergency.
Horticulture: Can be transplanted in sand early in the season before native conditions become too dry.
Note: The plant reproduces most readily by rhizomes.
Similar Species: *Elymus mollis* has its spikelets arranged in a single spike at the top of the culm.

References: Griffin, D. and I.G. MacQuarrie, 1980. The living sands, coastal dunes of P.E.I. *Nature Canada*, vol. 9 (2), 42–6.

Giant bulrush

Latin Name: *Scirpus acutus*
Genus: A cosmopolitan genus of nearly 200 species; 16 in Atlantic Canada
Family: Cyperaceae (Sedge)
Etymology: Latin name of the bulrush; acute
Habitat: In fresh or brackish water near the shores of ponds and lakes
Range: Nova Scotia and Prince Edward Island to Alaska, south to North Carolina and California; Europe
Longevity: Perennial
Flowering: August
Features: Simple flowers are grouped in several to many spikelets and are borne in axils of long scales, which are red-dotted and completely embrace the achene. The firm, straight stem is round in cross-section.
Fruit: Achene

Height: Up to 2 m.
Ecology: Nutlets and tubers are good duck food, while geese prefer root-stocks and herbage. This plant provides support for insects and shelter for young fish.
Uses: Roots and white stem bases may be eaten all year round, while the pollen can be used in making cakes, and the seeds are also edible. Root-stocks can be dried and beaten for use as a breadstuff, and young roots may be boiled to produce a sweet syrup.
Note: Sometimes classified as *S. lacustris*
Similar Species: *S. validus* has short spikelet scales that expose the achene.

References: Terrell, C.B., 1930. *Wild fowl and fish attractions for South Dakota*. Game and Fish Commission, Pierre, South Dakota.

Hare's-tail
also called cottongrass

Latin Name: Eriophorum spissum
Genus: About 20 species of the north temperate and Arctic zones; 7 in Atlantic Canada
Family: Cyperaceae (Sedge)
Etymology: Greek *erion*, wool or cotton, and *phoros*, bearing; crowded
Habitat: Dryish bogs, swales, and muskeg
Range: Labrador to Alaska, south to Pennsylvania and Wisconsin
Longevity: Perennial
Flowering: Late May
Features: The white bristles of the flowers become cotton-like as the plant matures with solitary terminal spikelets. The leaves are reduced to a bladeless sheath.
Fruit: Triangular achene
Height: 15–70 cm.
Note: Until pollination has occurred, this plant could be mistaken for almost any other sedge but is easily recognized once the cotton-like seed heads mature.
Similar Species: Forma *erubescens* has reddish-brown bristles. *E. Chamissonis* has a solitary stem.

Sedge

Latin Name: *Carex* spp.
Genus: Approximately 793 species throughout the world; 100 in Atlantic Canada
Family: Cyperaceae (Sedge)
Etymology: Greek *keirein*, to cut (alluding to sharp leaves)
Habitat: Many species prefer swamps, swales, damp meadows, and roadside ditches, while some are found in moist woodlands.
Range: Many species occur throughout Canada and into the United States, while others have a more restricted range.
Longevity: Perennial
Flowering: Early to late summer
Features: Grass-like plants with stems, which are triangular and have unisexual flowers. Stamens and pistils are in separate flowers, which may be in separate spikes, in different parts of the same spike, or scarcely distinguishable within a spike.
Fruit: Achene enclosed in a perigynium
Height: Up to 80 cm.
Ecology: This group of plants provides food for waterfowl, moose, beaver, deer, and muskrat.
Note: Mature perigynia are necessary for the identification of almost all species. This genus has a large number of varieties and forms.
Most Common Species: *Carex stipata, C. trisperma, C. scoparia, C. crinita, C. Pseudo-Cyperus, C. lacustris,* and *C. lurida*

References: Kubichek, W.F., 1933. Report on the food of five of our most important game ducks. *Iowa State College Journal of Science*, vol. 8 (1), 107–26.

McAtee, W.L., 1939. *Wildfowl food plants*. Collegiate Press, Inc., Ames, Iowa.

Jack-in-the-pulpit
also called Indian turnip

Latin Name: Arisaema Stewardsonii
Genus: About 100 species (most occur in Asia); 1 species in Atlantic Canada
Family: Araceae (Arum)
Etymology: Greek *aris*, arum, *haima*, blood (some species have spotted leaves); named for one of its discoverers, Stewardson Brown
Habitat: Rich, low woods, or along the edges of intervals and mucky areas in thickets
Range: Nova Scotia and Prince Edward Island to Minnesota, south to New Jersey and North Carolina
Longevity: Perennial
Flowering: Early June
Features: The hooded spathe, longitudinally stipped with white, protects a spadix whose upper part does not flower. The large compound leaves have 3 leaflets.
Fruit: Red berries clustered on lower part of spadix
Height: 30–100 cm.

Uses: The dry roots may be ground into starchy flour after drying and roasting. When fresh, it burns the mouth. Indians boiled and ate the berries.
Horticulture: The plant requires moist, rich soil and partial shade.
Note: Like most spring wildflowers, this one should not be picked indiscriminately.
Similar Species: Often classified as *A. triphyllum*; *Sparganium eurycarpum* has a rough, burr-like flowering head.

References: Bierzychudek, P., 1982. The demography of jack-in-the-pulpit, a forest perennial that changes sex (*Arisaema triphyllum*, New York). *Ecological Monographs*, vol. 52 (4), 335–51.

Doust, J.L. and P.B. Cavers, 1982. Sex and gender dynamics in jack-in-the-pulpit, *Arisaema triphyllum* (Araceae). *Ecology*, vol. 63 (3), 797–808.

Ellis, T.T. and V.E. Wiedeman, 1976. Description and breaking of dormancy in corms of jack-in-the-pulpit *Arisaema* spp. *Transactions Kentucky Academy of Science*, vol. 37 (3–4), 68–71.

Water arum

Latin Name: *Calla palustris*
Genus: There is only one species in this genus, and it occurs in Atlantic Canada.
Family: Araceae (Arum)
Etymology: Name of unknown meaning used by Pliny; of marshes
Habitat: Rooted in calm shallow water in bogs, ponds, and lake and stream edges
Range: Newfoundland to Alaska, south to New Jersey, Wisconsin and Minnesota; Eurasia
Longevity: Perennial
Flowering: June 15–July
Features: The flowers are in a thick fleshy spadix, partly enclosed or subtended by a milky white spathe. The glossy green leaves are heart-shaped.
Fruit: Red berries
Height: 10–30 cm.
Ecology: Rhizomes are reported to be eaten by moose.
Uses: When taken in spring before the leaves come out, the roots can be dried in an oven after removing the fibrous parts. After grinding, they are boiled and left standing for a few days and can be mixed with other flour and baked into bread.
Note: The Arum family is composed of a few species of distinctive plants.

Sweet flag
also called calamus

Latin Name: *Acorus Calamus*
Genus: 2 species, one of which grows in Atlantic Canada
Family: Araceae (Arum)
Etymology: Greek *akoras*, name of a plant with aromatic roots; Latin *calamus*, old name of a reed
Habitat: Marshes, along rivers, shallow pool edges, and wet meadows where the bases of the plants are continually submerged
Range: Nova Scotia and Prince Edward Island to Oregon, south to Florida and Texas; introduced into Europe
Longevity: Perennial
Flowering: July–August
Features: The cylindrical spadix is borne on the side of the two-edged stem and has a spathe that appears like a continuation of the stem.
Fruit: A dense head of a few brown, dry seeds

Height: To 2 m.
Uses: In spring, the undeveloped leaves in the centre of the stalk may be eaten raw. The fresh root is peppery with a peculiar soapy taste, but can be candied. To candy, boil the root two or three days, then cut up and boil a few minutes in sugared water.
Horticulture: Moist rich soil; ample sunlight
Note: The plant is aromatic, especially the thick creeping rhizome.
Similar Species: *Iris versicolor* has a showy blue flower. See also *Sparganium*.

References: Rizvi, S.J.H., 1981. Discovery of natural herbicides (*Acorus calamus, Azadirachta indica* (neem) and *Ocimum sanctum*). *Pesticides*, vol. 15 (3), 40–42.

Rost, L.C.M., 1979. Biosystematic investigations with *Acorus* (calamus), 4. A synthetic approach to the classification of the genus. *Planta Medica*, vol. 37 (4), 289–307.

Lesser duckweed

Latin Name: *Lemna minor*
Genus: About 5 species in the northern hemisphere; 2 in Atlantic Canada.
Family: Lemnaceae (Duckweed)
Etymology: Name of an aquatic plant mentioned by Theophrastus; smaller
Habitat: Floating on ponds and stagnant pools
Range: Throughout the world, except in the colder regions
Longevity: Perennial
Flowering: Often found until late October
Features: Monoecious flowers consist of a single anther or pistil, but are usually not present. The minute flattened plants are oval, not differentiated into stems and leaves, and have no vascular tissue in the single root.

Fruit: Tiny, one-seeded capsule
Ecology: Used as food by wild ducks
Note: Reproduction is by budding or by bulblets, which sink in autumn and rise to the surface in spring. Duckweed is thought to be the smallest and simplest flowering plant in the world.
Similar Species: *Spirodela polyrhiza* has several roots and a reddish under-surface.

References: Daubs, E.H., 1966. A monograph of *Lemnaceae*. *Illinois Biological Monographs*, vol. 34, 1–118.

Gustine, D.L., 1969. *Lemna minor* L., Ph.D. thesis, Michigan State University.

Severi, A. and R.B. Fornasiero, 1983. Morphological variations in *Lemna minor* L. and possible relationships with abscisic acid. *Caryologia*, vol. 36 (1), 57–64.

Blue-bead lily
also called clintonia or corn-lily

Latin Name: Clintonia borealis
Genus: 6 species, 1 of which is in Atlantic Canada
Family: Liliaceae (Lily)
Etymology: Dedicated to DeWitt Clinton (former governor of New York); northern
Habitat: Deciduous or mixed woodlands
Range: Labrador to Manitoba, south to Georgia
Longevity: Perennial
Flowering: Early June
Features: Its yellow flowers have 3 petals and 3 sepals and are grouped 3–6 in an umbel. The parallel-veined leaves are oval to elliptical and wrap around each other at the base of the plant.
Fruit: Dark blue berries
Height: 10–40 cm.
Ecology: Deer eat the young leaves

Uses: Very young unfurling leaves taste like cucumber in salads or are good when boiled as greens, but become bitter when older. Poisonous berries. Hunters used to rub their traps with the roots because bears are attracted to the odor.
Horticulture: These lilies are not very suitable for transplanting, but sometimes survive in relatively acidic soil and full shade.
Note: One of the most distinctive and best known of our spring flora
Similar Species: Erthronium americanum has a solitary flower and mottled leaves. See *Cypripedium* spp.

References: Plowright, R.C., 1981. Nectar production in the boreal forest lily, *Clintonia borealis. Canadian Journal of Botany*, vol. 59 (2), 156–60.

False Solomon's seal

Latin Name: *Smilacina racemosa*
Genus: About 20 species in boreal and northern temperate regions of east Asia and North America; 3 in Atlantic Canada
Family: Liliaceae (Lily)
Etymology: A diminutive of *Smilax*, the name used by Tournefort (1700); *racemus*, racemed
Habitat: Open deciduous woods, along the edges of thickets and in clearings
Range: Prince Edward Island and Nova Scotia to British Columbia, south to Tennessee and Arizona
Longevity: Perennial
Flowering: May–July
Features: Numerous white flowers with six-petalled perianth in a terminal inflorescence. The alternate, parallel-veined leaves are numerous and mostly sessile.
Fruit: Red berry, dotted with purple
Height: Up to 90 cm.
Ecology: Berries eaten by some birds and mice
Uses: Youngshoots make a passable asparagus; root-stocks can be pickled; berries are somewhat palatable but are cathartic; and the root-stocks can be soaked in lye and parboiled as a vegetable.
Horticulture: At least partial shade and moist sandy loam with medium pH (5.5–6.5)
Note: Comparative scarcity; should not be used when other vegetable food is available
Similar Species: Forma *foliosa* has lower branch of the panicle in upper leaf axil. *Streptopus* spp. has single or paired flowers.

Three-leaved false Solomon's seal

Latin Name: Smilacina trifolia
Genus: About 20 species (boreal and northern temperate regions of eastern Asia and North America); 3 in Atlantic Canada
Family: Liliaceae (Lily)
Etymology: A diminutive of *Smilax*, name used by Tournefort (1700) for these plants; three-leaved
Habitat: Sphagnum bogs, wet meadows, and swamps
Range: Labrador to British Columbia, south to New Jersey and southern Alberta; Siberia
Longevity: Perennial

Flowering: June–July 15
Features: The plant has a few white flowers on a stalked raceme and 2–4 leaves tapering to a sheathing base.
Fruit: Dark red berry
Height: Up to 50 cm.
Horticulture: Cool, wet, acidic soil
Note: This is the smallest of the solomon seals.
Similar Species: *S. stellata* has a zigzag stem with more leaves and striped black or green berry. *Maianthemum canadense* does not have six-parted flowers.

Starry false Solomon's seal

Latin Name: *Smilacina stellata*
Genus: About 20 species, in eastern Asia and North America; 3 in Atlantic Canada
Family: Liliceae (Lily)
Etymology: Diminutive of *Smilax*, name used by Tournefort (1700); starry
Habitat: Around the coast, in marshes and wet meadows
Range: Labrador to British Columbia, south to Virginia and California; Europe
Longevity: Perennial
Flowering: Early July
Features: The white flowers have 6 perianth parts and occur a few in a raceme that is sessile, or nearly so. The 7–12 leaves are alternate, glaucous, broad, and sub-clasping at their base.

Fruit: Red berry
Height: 20–50 cm.
Ecology: Berries are eaten by some birds and mice.
Uses: The plant can be used when young as asparagus. Before flowering, the young stems and leaves are as palatable as dandelion greens.
Horticulture: Rich, slightly acidic, humus soil; shade
Similar Species: *S. trifolia* has 3 leaves.

References: Galway, D.H., 1945. The North American species of *Smilacina*. *American Midland Naturalist*, vol. 33, 644–66.

Wild lily of the valley

Latin Name: Maianthemum canadense
Genus: 3 species in the temperate northern hemisphere; 1 in Atlantic Canada.
Family: Liliaceae (Lily)
Etymology: Latin *Maius*, May, Greek *anthemon*, flower; Canadian
Habitat: Common in a wide variety of habitats, but one of the first plants to appear under conifers
Range: Labrador to Minnesota, south to mountains of North Carolina and Georgia
Longevity: Perennial
Flowering: Early June
Features: Numerous white flowers with four-parted perianth occur in a terminal inflorescence. The leaves, usually 2, are alternate on the stem and have a heart-shaped base.
Fruit: Red berry
Height: 5–20 cm.
Ecology: Berries are eaten by birds, chipmunks, mice, and hares
Uses: The berries, with a not-unpalatable taste, are somewhat cathartic and should be eaten with caution.
Note: A colonial species, spreading by rhizomes
Similar Species: Smilacina trifolia, found in wet areas and has floral parts in three's. *Convallaria majalis* is an introduced species with united perianth parts.

Rose twisted-stalk
also called liverberry

Latin Name: *Streptopus roseus*, var. *perspectus*
Genus: 7 species in North America and east Asia; 2 occur in Atlantic Canada.
Family: Liliaceae (Lily)
Etymology: Greek *streptos*, twisted, *pous*, foot or stalk; rose-coloured
Habitat: Acid soils, thickets, coniferous, and mixed woods
Range: Labrador and Newfoundland to Michigan, south to Pennsylvania and North Carolina
Longevity: Perennial
Flowering: June
Features: The numerous bell-like, rose–purple flowers are borne singly on a jointed stalk. Lanceolate leaves occur alternately on the unbranched, arching stem and have parallel veins.
Fruit: Red berry
Height: 20–60 cm.
Uses: The berries are edible but cathartic and should be used with caution. Leaves and young shoots may be eaten in a salad, and creamed soup can be made from the shoots.
Horticulture: Shaded, well-drained, rich soil
Similar Species: *S. amplexifloius*, var. *americanus*, has greenish-white flowers.

Lily of the valley

Latin Name: *Convallaria majalis*
Genus: A monotypic genus
Family: Liliaceae (Lily)
Etymology: Latin *convallis*, a valley; blooming in May
Habitat: A garden plant that occasionally persists or spreads in patches near old houses, cemeteries, or occasionally along roadsides
Range: Introduced and naturalized from Europe
Longevity: Perennial
Flowering: May
Features: The white flowers have a six-parted perianth and occur several in a one-sided raceme. Its basal leaves are oval to elliptical.
Fruit: Red berry
Height: 10–20 cm.

Uses: The leaves and flowers contain a toxin that causes irregular heartbeat and pulse, usually accompanied by confusion and digestive upset.
Note: A colonial species that spreads by stolons
Similar Species: *Maianthemum canadense* has a four-parted perianth, with heart-shaped leaves at the base.

References: Cannon, R., 1982. Fragrant flowers (*Dianthus gratianopolitanus, Narcissus jonquilla, Convallaria majalis, Paeonia lactiflora, Malus baccata, Nicotiana alata*). American Horticulturalist, vol. 61 (2), 26–7.

Chapman, D., 1981. Ground covers adorn surfaces in shady, low maintenance areas (Including *Convallaria majalis, Hedera helix, Hosta* species). *Weeds, Trees and Turf,* vol. 20 (6), 26–8.

Indian cucumber root

Latin Name: *Medeola Virginiana*
Genus: Only 1 species, occurring in Atlantic Canada
Family: Liliaceae (Lily)
Etymology: Named for the sorceress *Medea* (for its imagined medicinal powers); of Virginia
Habitat: Rich, open, deciduous woodlands
Range: Prince Edward Island and Nova Scotia to western Ontario and Minnesota, south to Florida and the Great Lakes states
Longevity: Perennial
Flowering: June

Features: Straw-coloured flowers with 3 recurved petals occur at the top of the stem and are subtended by a whorl of 3 or more small leaves. A circle of 5–7 elongated leaves occurs at two-thirds of stem height.
Fruit: Black or purplish berries
Height: 20–90 cm.
Uses: The root-stock is crisp and starchy, with a delicate taste of cucumber, and can be nibbled plain or dressed with vinegar and oil.
Horticulture: Moist, rich soil (pH 5.5–6.5); shaded

Nodding trillium

Latin Name: *Trillium cernuum*
Genus: About 25 species in temperate northern hemisphere; 3 in Atlantic Canada
Family: Liliaceae (Lily)
Etymology: Latin *tres*, three (all floral parts in three's; nodding
Habitat: Alluvial soils, flood plains, and deciduous climax forest
Range: Newfoundland to the Mackenzie Valley to Wisconsin, south to Georgia, Iowa, and Tennessee
Longevity: Perennial
Flowering: May 20–June 15
Features: Pale pink, sweet-scented, three-parted flowers hang under the 3 leaves on a short stalk. Its leaves occur in a single whorl and taper to a sessile base.
Fruit: Dark red berry
Height: 15–60 cm.
Ecology: Berries eaten by squirrels

Uses: Trillium roots are highly emetic and the berries are open to suspicion. The young, unfolding plants can be eaten as greens, but this is not encouraged because of the scarcity of the plant.
Horticulture: Moist, sandy loam with mulch, medium pH (6.0–7.0), and light shade
Note: Most northerly distributed of the trilliums and the only member we have of the Lily family without parallel-veined leaves. The plant may die if picked before the leaves manufacture enough food for storage in the root for next spring's growth.
Similar Species: *T. erectum* has purple–brown petals. *T. undulatum* has white petals with reddish-pink bases.

References: Gates, R.R., 1917. A systematic study of the North American genus *Trillium*, its variability, and its relation to *Paris* and *Medeola*. Annals of the Missouri Botanical Garden, vol. 4, 43–92.

Painted trillium

Latin Name: *Trillium undulatum*
Genus: 25 species in the temperate northern hemisphere; 3 occur in Atlantic Canada.
Family: Liliaceae (Lily)
Etymology: Latin *tres*, three (floral parts are in three's); waxy
Habitat: Intervales and open, dry, to rather rich woods
Range: Nova Scotia to Manitoba, south to Georgia
Longevity: Perennial
Flowering: May 20 – June 20
Features: The flower has 3 white petals with pink-striped bases; other floral parts and leaves occur in three's.
Fruit: Red berry
Height: 10–50 cm.
Uses: The young unfolding plants can be eaten as greens, but the berries are open to suspicion. Trillium roots are reported to be highly emetic and should be avoided.
Horticulture: Well-drained humus soil; partial shade
Note: The plant may die if picked before the leaves have a chance to manufacture enough food for storage in the root for next spring's growth.
Similar Species: *T. erectum* has dark purple flowers; *T. cernuum* has pale pink flowers recurved under the leaves.

References: Cruise, J.E. and L. Gad, 1974. Trilliums—their unusual forms. *Ontario Naturalist*, vol. 14(1), 33-6.

Blue-eyed grass

Latin Name: Sisyrinchium monatnum, var. crebrum
Genus: About 75 species in western hemisphere; 4 in Atlantic Canada
Family: Iridaceae (Iris)
Etymology: Sisyrinchium, name used by Theophrastus for another plant; of the mountains
Habitat: Fields, meadows, roadsides, and open fields
Range: Nova Scotia to Alberta and Minnesota, south to West Virginia
Longevity: Perennial

Flowering: Late May – June
Features: Blue-violet flowers have yellow centres, six-parted perianth, occur in clusters of 1 to several, and last only 1 day. The winged stem has grass-like leaves and a spathe.
Fruit: Capsule
Height: 10–60 cm.
Horticulture: Moist, well-drained soil with medium pH (6.0–7.0); full sunlight
Note: Flowers open only in bright sunlight.
Similar Species: S. angustifolium and S. atlanticum generally have 2 or more spathes and flowers peduncled from axil of leafy bract.

Blue flag

Latin Name: Iris versicolor
Genus: 200 species in the northern hemisphere; 4 in Atlantic Canada
Family: Iridaceae (Iris)
Etymology: Greek *Iris*, rainbow; variously coloured
Habitat: Meadows, swamps, along streams, and in wet pastures
Range: Labrador to Manitoba, south to West Virginia
Longevity: Perennial
Flowering: June–July
Features: The large purple-blue flowers are prominently veined and have flat petals that are half as long as the sepals. Its leaves are flattened, equitant, and up to 3 cm. wide.
Fruit: Flattened capsule
Height: 20–80 cm.
Horticulture: Moist soil with a pH 6.0–7.0, full sunlight
Note: All parts are toxic and reported to be violently emetic and cathartic.
Similar Species: The white-flowered forma *Murrayana* is rare; *I. prismatica* is much more slender as is *I. Hookeri*. It also resembles *Acorus Calamus*, which is found in shallow water.

Stemless Lady's slipper
also called common Lady's slipper and moccasin flower

Latin Name: Cypripedium acaule
Genus: About 20 in total; 5 are in Atlantic Canada
Family: Orchidaceae (Orchid)
Etymology: Name incorrectly taken from *Cypris*, Venus, and *pedilon*, shoe; stemless
Habitat: In acidic soil in dry or moist woods
Range: Newfoundland to northern Alberta, south in the mountains to North Carolina and Tennessee
Longevity: Perennial
Flowering: June–early July
Features: A showy solitary flower with a pink moccasin-shaped lip. It has only 2 leaves at the base of the naked floral scape.
Fruit: Capsule with minute seeds
Height: 10–55 cm.
Ecology: The flowers are shaped to attract insects for pollination.

Horticulture: Requires acidic soil but is very difficult to grow. Some gardeners have met with success in transplanting, but the plant often survives only a few seasons.
Note: Floral emblem of Prince Edward Island. Two-flowered forms occur rarely.
Similar Species: Forma *albiflorum* has white flowers, *C. Calceolus* has yellow flowers, and *C. reginae* has a white floral lip flushed with purple.

References: Plowright, D.C., J.D. Thomson, and G.R. Thaler, 1980. Pollen removal in *Cypripedium acaule (Orchidaceae)* in relation to aerial fenitrothion spraying in New Brunswick (Bumble bee pollinators, pesticide toxicity). *The Canadian Entomologist*, vol. 112(8), 765–9.

Stoutmire, W.P., 1967. Flower biology of the lady's slippers *(Orchidacea: Cypripedium)*. *Michigan Botanist*, vol. 6(4), 159–75.

Ragged fringed orchid

Latin Name: *Habenaria lacera*
Genus: 400 species, nearly world-wide; 13 in Atlantic Canada
Family: Orchidaceae (Orchid)
Etymology: Latin *habena*, a thong or rein (alluding to shape of floral lip on some species); lacerated
Habitat: Wet places such as bogs, meadows, damp fields, dune hollows, and poorly drained clay soils
Range: Newfoundland to Michigan, south to Alabama and Texas
Longevity: Perennials
Flowering: July–August
Features: Greenish- or yellowish-white flowers grow on a spike, have 1 fertile anther, 3-divisional and deeply fringed lip, and a spur. The erect stem is unbranched.
Fruit: Three-valved capsule
Height: 20–80 cm.
Note: Flowers are adapted to attract insects for pollination. Hybridizes with *H. psycodes* and others
Similar Species: *H. blephariglottis* is smaller and has white flowers.

Curled dock

Latin Name: Rumex crispus
Genus: About 150 species, chiefly in the temperate zones; 12 in Atlantic Canada
Family: Polygonaceae (Buckwheat)
Etymology: The ancient Latin name; curled
Habitat: Meadows, pastures, waste places, and along roadsides
Range: Throughout temperate North America; introduced from Eurasia
Longevity: Perennial
Flowering: Mid June – July
Features: Small greenish flowers are whorled in a branched inflorescence but with no corolla: a coarse herb with lanceolate, crisped, and undulate leaves
Fruit: Triangular achene
Height: To 90 cm.
Uses: It was traditionally collected, boiled, and given to girls in Newfoundland for its iron content. Root can be used as a tonic, astringent, a gentle laxative, or tea. The leaves are high in vitamin C and can be used as a spring green or pot-herb. Flour can be made from the seeds.
Similar Species: R. orbiculatus has broad, flat leaves.

***

References: Bentley, S. and J.B. Whittaker, 1979. Effects of grazing by a chrysomelid beetle, *Gastrophysa viridula*, on competition between *Rumex obtusifolius* and *Rumex crispus*. *Journal of Ecology*, vol. 67(1), 79–80.

Hume, L. and P.B. Cavers, 1981. A methodoligical problem in genecology. Seeds versus clones (of *Rumus crispus* from two contrasting habitats) as source material for uniform gardens. *Canadian Journal of Botany*, vol. 59(5), 763–8.

Hume, L. and P.B. Cavers, 1982. Geographic variation in a widespread perennial weed, *Rumex crispus*. The relative amounts of genetic and environmentally induced variation among populations of North America. *Canadian Journal of Botany*, vol. 60(10), 1928–37.

Sheep-sorrel

Latin Name: Rumex Acetosella
Genus: About 150 species, most abundant in the temperate zones; 12 in Atlantic Canada
Family: Polygonaceae (Buckwheat)
Etymology: The ancient Latin name; old generic name, meaning little sorrel
Habitat: Meadows, pastures, roadsides, burnt lands, barrens, and in worn-out fields on sour soils
Range: In every province and throughout North America; introduced from Eurasia
Longevity: Perennial
Flowering: June–October
Features: Small green flowers tinged with red; male and female flowers are on separate plants. It has arrow-shaped leaves.
Fruit: Achene
Height: To 40 cm.
Uses: Excellent for seasoning in a salad; leaves can be cooked as greens, used as a pot-herb, or eaten fresh as a nibble. American Indians are reported to have used the seeds in preparing meals.
Note: Difficult to eradicate, except by sweetening soil with lime. The acidic taste is from potassium oxalate.
Similar Species: R. acetosa, a larger plant, does not have flaring basal leaf lobes.

Orach

Latin Name: *Atriplex patula*
Genus: About 150 species in western and eastern hemispheres; 2 in Atlantic Canada
Family: Chenopodiaceac (Goosefoot)
Etymology: The ancient Latin name; typical
Habitat: Around the coast, salt marshes, flooded dike lands, and upper regions of beaches
Range: Newfoundland to British Columbia, south to South Carolina and California
Longevity: Annual
Flowering: July–September
Features: Separate staminate and pistillate flowers occur, the latter without a calyx and enclosed within 2 broad bracteoles. The green leaves are scarcely mealy, but sometimes greyish when young and often hastate.
Fruit: One-seeded utricle
Height: Up to 1 m.
Note: This is a highly variable species, with the different varieties tending to intergrade in appearance.
Similar Species: Var. *hastata* has basal lobes on all leaves, var. *patula* has lanceolate to oblong leaves, and var. *oblanceolata* has oblanceolate leaves. See *Chenopodium*, which has alternate leaves.

Spring-beauty

Latin Name: *Claytonia caroliniana*
Genus: About 20 species of North America; 2 in Atlantic Canada
Family: Portulacaceae (Purslane)
Etymology: Named for John Clayton, an early American botanist; of Carolina
Habitat: Rich hardwood forests
Range: Newfoundland to Ontario, south to the mountains of Tennessee and North Carolina
Longevity: Perennial
Flowering: May 20 – June 15
Features: Loose raceme of white or pale pink flowers, which have 5 rose-veined petals and 2 sepals. Only 2 succulent leaves occur opposite each other and have distinct petioles.
Fruit: Capsule
Height: Up to 30 cm.
Ecology: Tubers are eaten by some small mammals.
Horticulture: Moist humus soil; shade to semi-shade
Similar Species: *C. fontana* has small white flowers.

References: Handel, S.N., 1978. New ant-dispersed species in the genera *Carex, Luzula,* and *Claytonia*. *Canadian Journal of Botany,* vol. 56(22) 2925–7.

Sand-spurry

Latin Name: *Spergularia rubra*
Genus: About 50 species in the temperate zone; 3 in Atlantic Canada
Family: Caryophyllaceae (Pink family)
Etymology: Name a derivative of *Spergula*; red
Habitat: Sandy places, gravelly soil in farmyards, waste places, and towns
Range: Newfoundland to Minnesota, south to Virginia; Vancouver Island to California; introduced from Europe.

Longevity: Annual or short-lived perennial
Flowering: June–September
Features: Small bright pink flowers have 5 petals, 5 sepals, and 3 styles. They open only on bright sunny days. The small linear leaves are opposite.
Fruit: Capsule
Height: 5–15 cm., forming mats

Stitchwort

Latin Name: *Stellaria graminea*
Genus: Nearly 100 species, chiefly in the northern temperate zone; 7 in Atlantic Canada
Family: Caryophyllaceae (Pink)
Etymology: Latin *stella*, a star (referring to flower); grass-like
Habitat: Grainfields, row crops, pastures, gardens, lawns, and waste places
Range: Newfoundland to Minnesota, south to Maryland and Missouri
Longevity: Annual
Flowering: Throughout summer
Features: Flowers have 5 white petals deeply divided to the base and longer than the sepals. Leaves are lanceolate, widest below the middle, and essentially sessile on the smooth stem.
Fruit: Capsule
Height: 20–50 cm.
Similar Species: *S. media* has ovate leaves with distinct petioles on the middle and lower ones.

References: Batra, S.W.T., 1979. Insects associated with weeds in the northeastern United States. III. Chickweed, *Stellaria media*, and stitchwort, *Stellaria graminea* (Carylophyllaceae). *Journal New York Entomological Society*, vol. 87(3), 223–35.

Lincoln, W.C. Jr., 1980. Laboratory germination of *Stellaria graminea*. *The Newsletter of the Association of Official Seed Analysts*, vol. 54(1), 73–4.

Field chickweed

Latin Name: *Cerastium arvense*
Genus: Nearly 100 species, mostly in temperate climates; 2 in Atlantic Canada
Family: Caryophyllaceae (Pink)
Etymology: Greek *cerastes*, horned (referring to recurved capsule); typical
Habitat: Fields and meadows
Range: Throughout Canada; Europe
Longevity: Perennial
Flowering: Early June
Features: The 5 showy white petals are two-lobed and are longer than the sepals. The linear or linear-lanceolate leaves are opposite on the stems, covered with downward-pointing hairs.
Fruit: Long capsule
Height: 10–25 cm.
Ecology: Seeds are used as food by songbirds.
Note: Many variations occur in this species.
Similar Species: *C. vulgatum* has smaller petals and hairy, stalkless leaves. *Stellaria* species are less hairy and have only 3 styles.

Water-lily

Latin Name: *Nymphaea odorata*
Genus: About 40 species, widely dispersed in the tropics and temperate regions; 1 in Atlantic Canada
Family: Nymphaeaceae (Water-lily)
Etymology: Ancient name from water-nymphs; fragrant
Habitat: Bog pools, lake margins, shallow muddy ponds, and slow-flowing rivers
Range: Newfoundland to Manitoba, south to Florida and Louisiana
Longevity: Perennial
Flowering: June–September
Features: Aquatic plant with large floating leaves and large white or pinkish and strongly fragrant flowers with numerous yellow stamens. The leaves are borne on the summit of a thick, spongy root-stock.
Fruit: 12–35 locular globule
Height: 50–100 cm.
Ecology: The plants are eaten by many animals, including beaver and muskrat; ducks are known to eat the seeds.
Uses: The young leaves and unopened flower buds are edible. Root-stocks can be used as a starchy vegetable; the seeds may be fried, used in soups, or popped like corn.
Horticulture: Easily transported into small ponds with a muddy bottom and exposed to full sunlight.
Note: The stomata are on the upper side of the leaf.
Similar Species: Var. *rosea* has smaller, pinkish petals. It is suspected of being an ecological form occurring when plants grow in dry conditions.

References: Masters, C.O., 1974. *Encyclopedia of the Water-Lily*. Neptune City, New Jersey.

Schneider, E.L. and T. Chaney, 1981. The flora biology of *Nymphaea odorata* (Nymphaeaceae). *The Southwestern Naturalist*, vol. 26(2), 159–65.

Buttercup
also called crowfoot

Latin Name: *Ranunculus* spp.
Genus: About 300 species in north temperate and Arctic regions; 12 in Atlantic Canada
Family: Ranunculaceae (Buttercup)
Etymology: Latin name applied by Pliny and meaning little frog, a reference to the wet habitat of many species
Habitat: Aquatic buttercups occur in slow-moving streams and shallow pools, while other species are found in salt marshes, intervales, wet woods, meadows, and pastures.
Range: Most species found throughout Canada and into the United States
Longevity: Perennial
Flowering: Late June – August
Features: Showy yellow flowers have 5, often waxy, petals. All species are herbaceous and have mostly basal leaves.

Fruit: Many achenes
Height: Up to 1 m.
Ecology: Used as food by waterfowl and upland game birds
Note: Buttercups have a bitter juice, which discourages grazing, but which is volatile and is dispelled when hay is cured.
Most Common Species: *Ranunculus Cymbalaria* occurs in salt marshes; *R. acris* is common in heavy or moist soil; *R. recurvatus* is in rich woodlands.

References: Gross, A.O., 1937. Food of the ruffed grouse. *Game Breeder and Sportsman,* vol. 41, 142-4.

Sarukhan, J., 1974. Studies on plants demography: *Ranunculus repens* L., *R. bulbosus* L. and *R. acris. Journal of Ecology,* vol. 62, 151-77.

Meadow rue

Latin Name: *Thalictrum polygamum*
Genus: About 120 species, mostly in northern hemisphere; 2 in Atlantic Canada
Family: Ranunculaceae (Crowfoot)
Etymology: A name of an unidentified plant mentioned by Dioseorides; polygamous
Habitat: Marshes, meadows, ditches, thickets and sometimes in flood plain climax forest.
Range: Labrador, south to Nova Scotia and northern New England
Longevity: Perennial
Flowering: July – early August
Features: Small white flowers in large clusters; some are perfect while others are unisexual. There is no corolla, but 4 or 5 sepals are present, which fall off early. Leaves are divided into many rounded leaflets with the terminal leaflet having 3 teeth.
Fruit: Many achenes
Height: 120–245 cm.
Horticulture: Moist, rich loam; sun or partial shade
Similar Species: Var. *hebecarpum* is less common and has a more compact inflorescence.

References: Melampy, M.N., 1981. Sex-linked niche differentiation in two species of *Thalictrum*. *American Midland Naturalist*, vol. 106(2), 325–34.

Yao, S. Shih-Yang, 1972. New alkaloids from *Thalictrum polygamum* Muhl. and *Berberis baluchistanica* Ahrendt. Ph.D. thesis, Pennsylvania State University, Ann Arbor, Michigan.

Marsh marigold

Latin Name: *Caltha palustris*
Genus: 15 species in Arctic and cool temperate areas in western and eastern hemispheres; 1 in Atlantic Canada
Family: Ranunculaceae (Crowfoot)
Etymology: *Caltha*, Latin name of a strong-smelling flower; of marshes
Habitat: Fresh water marshes
Range: Labrador to Alaska, south to South Carolina; Eurasia
Longevity: Perennial
Flowering: May and early June
Features: The yellow flowers have no true petals, only 5 showy sepals, but have numerous stamens and pistils. Some of the shallow-lobed, kidney-shaped leaves are on the smooth, hollow stem, but most are basal.

Fruit: Many-seeded follicle
Height: 20–60 cm.
Uses: The flower-buds can be pickled by soaking in salt and cooking in spiced vinegar. Young leaves are edible after boiling for thirty minutes to an hour in two changes of water.
Note: The raw parts and water in which they are boiled are toxic containing helleborin. This plant is a harbinger of spring.

References: Falinska, K., 1980. Experimental studies of the reproductive strategy of *Caltha palustris* L. populations. *Polish Journal of Ecology*, vol. 27(4), 527–43.

Peterson, R.L. and M.G. Scott, 1979. Some aspects of carpal structure in *Caltha palustris* L. *(Ranunculaceae)*. *American Journal of Botany*, vol. 66(3), 334–42.

Goldthread

Latin Name: Coptis trifolia
Genus: About 15 species of the north temperate and Arctic area; one in Atlantic Canada
Family: Ranunculaceae (Crowfoot)
Etymology: Greek *coptein*, to cut (refers to divided leaves); three-leaved
Habitat: Coniferous woods, swamps, bogs, and roadsides
Range: Labrador to Alaska, south to Maryland; in the mountains to North Carolina
Longevity: Perennial
Flowering: June–July
Features: The tiny white solitary flowers have 5–7 petals and deciduous sepals and occur on a scape. Evergreen basal leaves are divided into 3 leaflets and arise from the yellow root.
Fruit: Follicle
Height: 10 cm.
Uses: The boiled root can be used as a tonic, an antiseptic, and as a cure for stomach-ache. Roots were used by many tribes of Indians to prepare a remedy for sore eyes.
Horticulture: Rich, humic soil (pH 4.0–5.0); shade
Similar Species: Subspecies *groenlandica* has larger follicles and is often the designation for the North American species.

Red baneberry

Latin Name: *Actaea rubra*
Genus: About 7 species in northern hemisphere; 2 in Atlantic Canada
Family: Ranunculaceae (Crowfoot)
Etymology: *Actaea*, ancient Latin name of the elder, transferred to this genus; red
Habitat: Rich soil of hardwood forests and along edges of intervales
Range: Labrador to Alaska, south to New Jersey and Nebraska
Longevity: Perennial
Flowering: May 15 – May 30
Features: Small white flowers in a long-stalked raceme and on slender peduncles. The 2–3 compound leaves have coarsely toothed leaflets that are pubescent on the lower surface. As soon as a flower bud opens, the small white petal-like sepals fall away.
Fruit: Several seeded, red berries

Height: Up to 61 cm.
Ecology: Berries eaten by ruffed grouse and some mice
Horticulture: Rich shaded soil
Note: Mildly poisonous for humans, with disagreeable taste
Similar Species: Forma *neglecta* has pure white berries. *A. pachypoda* has white berries on thick stalks.

References: Favreau, G., Y. Raymond, and J. Masquelier, 1969. Study of red pigment isolated from the fruit of *Actaea rubra* Willd. *Nature Canada*, vol. 96(2), 191–202.

Raymond, Y., G. Favreau, and D. Grenier, 1970. Dosage of aluminum and potassium in different parts of *Actaea rubra* Willd. *Nature in Canada*, vol. 97(4), 489–90.

Raymond, Y., D. Grenier, F. Mercier, and P.P. Leblanc, 1971. Qualitative study of fatty acids of *Actaea rubra* Willd seeds. *Nature Canada*, vol. 98 (6), 955–8.

Yellow rocket
also called winter cress

Latin Name: *Barbarea vulgaris*
Genus: About 12 species of Eurasia and North American; one in Atlantic Canada
Family: Brassicaeceae (Mustard)
Etymology: Anciently called the Herb of St. Barbara; typical
Habitat: Hayfields, pastures, along roadsides and in intervals; introduced from Europe
Range: In all provinces and widely distributed in the United States
Longevity: Biennial or perennial
Flowering: Late May and June
Features: Large clusters of small yellow four-petalled flowers occur in racemes. The upper leaves are toothed and ovate, while the lower are lyre-shaped. Both often have clasping bases on the smooth stem.
Fruit: Silique
Height: To 60 cm.

Uses: The leaves are good as greens or in salads. Flower buds are also edible. Old capsules, after shedding seeds, can be collected with their stalks and used in dried-flower arrangements.
Note: One of the first mustards to flower each year. It was apparently introduced in grain or grass seed.
Similar Species: Var. *arcuata* has more lax and open pedicels. *Brassica Kaber* has thin, often hairy leaves.

References: Dutt, T.E., R.G. Harvey, and R.S. Fawcett, 1982. Feed quality of hay containing perennial broadleaf weeds. *Agronomy Journal*, vol. 74(4), 673–6.

Fawcett, R.S. and V.M. Jennings, 1978. Today's weed: yellow-rocket. *Weeds Today*, vol. 9(3), 21.

Morgan, D.D., 1971. A study of four weed hosts of *Sclerotipia* species in alfalfa fields. *Plant Disease Reporter*, vol. 55(12), 1087–9.

Sea-rocket

Latin Name: Cakile edentula
Genus: About 8 species in the north temperate zone and into tropical America; 1 in Atlantic Canada
Family: Brassicaceae (Mustard)
Etymology: The old Arabic name; without teeth
Habitat: Sandy beaches, dunes, cliffs, and on shingle beaches
Range: On the coast from Labrador to South Carolina; Pacific coast; and local about the Great Lakes

Longevity: Annual
Flowering: July–September
Features: The pale purple flowers have 4 petals. Fleshy, simple leaves occur on stems that may be much branched.
Fruit: Silique
Height: Up to 85 cm.
Uses: A few leaves give a lift to a salad when mixed with other greens. When steamed to be eaten as greens, they lose the strong horse-radish-like flavour.

Pitcher-plant

Latin Name: Sarracenia purpurea
Genus: About 10 species; 1 in Atlantic Canada
Family: Sarraceniaceae (Pitcher-plant)
Etymology: Named in honour of Michael Sarrasin de l'Étang (1659–1734); purple
Habitat: Bogs, bog meadows, and sphagnum lake margins
Range: Newfoundland to Alberta, south to Maryland and Illinois
Longevity: Perennial
Flowering: June 15–July
Features: The large solitary purple-red flower nods at the top of its stalk. Leaves occur in a basal rosette and are tubular-shaped to hold water in which insects become trapped when they cannot climb the smooth interior walls. The leaf lip has downward-pointing hairs, which also make it difficult for the insect to retrace its steps.
Fruit: Capsule
Height: 30–50 cm
Ecology: The plant digests insects to obtain nitrogen in a nitrogen-deficient environment. The pitcher-plant mosquito lays its eggs in water inside the leaves.
Uses: Indians are reputed to have made a potion from the plant to fight smallpox.
Horticulture: Peaty or sphagnum soil (pH 4.5–5.5); full sun
Note: This is the floral emblem of Newfoundland.

References: Cody, W.J. and S.S. Talbot, 1973. The pitcher plant, *Sarracenia purpurea* L. in the northwestern part of its range. *The Canadian Field-Naturalist,* vol. 87, 318–20.

Cruise, J.E. and P.M. Catling, 1971. The pitcher plant in Ontario. *Ontario Naturalist,* vol. 9(1), 18–21.

Savage, C. and A. Savage, 1979. Canada's carnivorous plants. *Nature Canada,* vol. 8(1), 4–12.

Round-leaved sundew

Latin Name: Drosera rotundifolia
Genus: About 100 species, mostly in Australia and South Africa; 2 are in Atlantic Canada
Family: Droseraceae (Sundew)
Etymology: Greek *droseros*, dewy; round-leaved
Habitat: Marshes, bogs, barrens, lake margins, ditches, and boggy black spruce woods
Range: Labrador to Alaska, south to Florida and California; Eurasia
Longevity: Perennial
Flowering: July 15 – August 15
Features: Small white, five-petalled flowers grow on a one-sided raceme that nods at its apex with the newest blossom always uppermost. Leaves are basal, rounded at tip, with red gland-bearing bristles.
Fruit: Capsule
Height: 5–30 cm.

Ecology: The sticky leaves capture and digest insects and small spiders to obtain nitrogen.
Uses: The leaves can be used as a substitute for rennet for curdling milk.
Note: The blossoms open one at a time and only in bright midday sunlight.
Similar Species: D. intermedia has narrow leaves and smooth petioles.

References: Hindley, K., 1980. The association of ladyslipper orchids and insectivorous plants. III. The association of *Cypripedium reginae* and *Cypripedium pubescens* with *Drosera rotundifolia* in bogs in Vermont. *The Orchid Digest*, vol. 44(6), 232-5.

Savage, C. and A. Savage, 1979. Canada's carnivorous plants. *Nature Canada*, vol. 8(1), 4-12.

Swales, D.E., 1975. An unusual habitat for *Drosera rotundifolia* L., its overwintering state and vegetative reproduction. *The Canadian Field-Naturalist*, vol. 89, 143-7.

Wild strawberry

Latin Name: *Fragaria virginiana*
Genus: 35 species in temperate northern hemisphere and Andean region; 3 occur in Atlantic Canada
Family: Rosaceae (Rose)
Etymology: Latin *fragra*, strawberry; Virginian
Habitat: Open woodlands, pastures, barrens, and fields
Range: Newfoundland to Alaska, south to Georgia and Oklahoma
Longevity: Perennial
Flowering: May – early June
Features: White five-petalled flowers with many stamens and pistils and on shorter stalks than the leaves; freely running plant with 3 rough-toothed leaflets
Fruit: Achenes in pits of red fruit
Height: 15 cm
Ecology: Many wild animals eat the fruit and leaves.

Uses: Leaves make a good herbal tea high in vitamin C. The fruit can be eaten fresh or used in shortcakes, tarts, sauces, jellies, jams, and preserves.
Horticulture: Well-drained humic soil (pH 5.5–6.5); in full sunlight
Similar Species: Var. *terrae-novae* has hairs on peduncles and petioles. *F. vesca* has achenes on unpitted fruit surface. See *Potentilla* species, which do not have berries in fruit.

References: Jurik, T.W., J.F. Chabot, and B.F. Chabot, 1982. Effects of light and nutrients on leaf size, CO_2 (carbon dioxide) exchange, and anatomy in wild strawberry *(Fragaria virginiana)*. *Plant Physiology*, vol. 70(4), 1044–8.

O'Neill, S.D., D.A. Priestley, and B.F. Chabot, 1981. Temperature and aging effects on leaf membranes of a cold hardy perennial, *(Fragaria virginiana)*. *Plant Physiology*, vol. 68(6), 1409–15.

Stault, G., 1963. Taxonomic studies in the genus *Fragaria*. *Canadian Journal of Botany*, vol. 40, 869–86.

Marsh cinquefoil

Latin Name: Potentilla palustris
Genus: 300 species in northern hemisphere; 15 in Atlantic Canada
Family: Rosaceae (Rose)
Etymology: A diminutive from Latin, *potens*, powerful, referring to reputed medicinal powers of one species; of marshes
Habitat: Undrained ponds, muddy shores, or in swamps above river estuaries
Range: Labrador to Alaska, south to New Jersey, Pennsylvania, and California; Eurasia
Longevity: Perennial

Flowering: June–August
Features: Flowers are several to many in a cyme and have 5 reddish-purple petals, which are shorter than the broad calyx lobes. Leaves are pinnately compound with 5–11 toothed leaflets.
Fruit: Mass of achenes on a dry receptacle
Height: 10–60 cm.
Note: The plants are variable, resulting in dispute over varieties.
Similar Species: P. pensylvanica has yellow petals and occurs in dry locations.

Rough-fruited cinquefoil
also called sulphur cinquefoil

Latin Name: *Potentilla recta*
Genus: About 300 species in the northern hemisphere; 15 in Atlantic Canada.
Family: Rosaceae (Rose)
Etymology: A diminutive from *potens*, powerful (originally applid to *P. anserina* for its once reputed medicinal powers); upright
Habitat: Pastures, hayfields, waste places, and along roadsides
Range: In all provinces, south to Tennessee and Kansas; introduced from Europe
Longevity: Perennial
Flowering: June 20 – July
Features: The sulphur-yellow five-petalled flowers occur in a cyme on an erect stalk and have numerous stamens. Leaves have 5–7 prominently toothed leaflets attached at one place.
Fruit: Mass of achenes on a dry receptacle
Height: 15–45 cm.
Note: Spreads rapidly by seeds and is becoming a nuisance in many areas.
Similar Species: Var. *sulphurea* has 7 leaflets on middle and lower stem leaves; var. *obscura* has 5 leaflets.

References: Batra, S.W.T., 1979. Insects associated with weeds in the northeastern United States. II. Cinquefoilis, *Potentialla norvegica*, and *Potentilla recta* (Rosaceae). *Journal New York Entomological Society*, vol. 87(3) 216-22.

Rough cinquefoil

Latin Name: *Potentilla norvegica*
Genus: About 300 species in northern hemisphere; 15 in Atlantic Canada
Family: Rosaceae (Rose)
Etymology: A diminutive of *potens*, powerful (referring to once reputed healing powers of one species); Norwegian
Habitat: Grainfields, hayfields, pastures, gardens, woods, and waste places
Range: Widespread in North America and Eurasia
Longevity: Biennial or short-lived perennial
Flowering: June–July
Features: Yellow five-petalled flowers have numerous stamens and pistils and occur in close, leafy clusters. The leaves are divided into 3 obovate leaflets that have a toothed margin.
Fruit: Mass of achenes
Height: 15–60 cm.
Note: This plant is highly variable and may be partly native and partly introduced.
Similar Species: *P. recta* has 5–9 leaflets and is taller.

References: Batra, S.W.T., 1979. Insects associated with weeds in the northeastern United States. II. Cinquefoilis, *Potentilla norvegica*, and *Potentilla recta* (Rosaceae). *Journal New York Entomological Society*, vol. 87(3) 216-22.

Werner, P.A. and J.D. Soule, 1976. The biology of Canadian weeds, 18: *Potentilla recta* L., *P. norvegia* L., *P. argentea* L. *Canadian Journal of Plant Science*, vol. 56, 561-603.

Silverweed

Latin Name: Potentilla anserina
Genus: About 300 species in the northern hemisphere; 15 in Atlantic Canada
Family: Rosaceae (Rose)
Etymology: A diminutive from *potens*, powerful, originally applied to this species from its once reputed medicinal powers; of geese
Habitat: Damp soil around coasts, salt marshes, and low dune areas
Range: Newfoundland to Alaska, south to New Jersey; inland around the Great Lakes and westward; Eurasia
Longevity: Perennial
Flowering: June–August
Features: Showy yellow flowers have 5 petals, 5 sepals, numerous stamens and pistils, and are borne singly on stalks. The basal leaves consist of 7–21 leaflets that are silky-silvery on the under-side.
Fruit: Mass of small achenes
Height: Prostrate
Uses: The thick, fleshy roots can be eaten either raw or cooked. They are especially good when candied. Its sweet potato-like flour can be enjoyed alone or in combination with almost anything else.
Note: Spreads by slender, many-jointed runners. The common name refers to the silvery under-side of the leaves.
Similar Species: Forma *sericea* has pubescence on both sides of leaves. Var. *Rolandii* occupies different habitats.

Queen of the meadow

Latin Name: Filipendula Ulmaria
Genus: About 11 species, mostly in northern temperate zone; 3 in Atlantic Canada.
Family: Rosaceae (Rose)
Etymology: Latin *filium*, a thread, and *pendulus*, hanging (allusion to roots of one species); old generic name (from fancied resemblance of leaflets to elm leaves)
Habitat: In low areas, around buildings, roadsides, waste places, neglected meadows, and field edges
Range: Newfoundland to Ontario; introduced from Europe

Longevity: Perennial
Flowering: Late July – August
Features: Large panicle of small creamy-white flowers. Plants are rather shrubby, with pinnately compound leaves and large leaflets that are white and woolly beneath.
Fruit: Achene
Height: 1–2 m.

References: Lindeman, A., P. Jounela-Ericksson, and M. Lounasmaa, 1982. The aroma composition of the flower of meadowsweet (*Filipendula ulmaria* [L.] Maxim.) (Possible food additives). *Food Science and Technology*, vol. 15(5), 286–9.

Avens

Latin Name: *Geum macrophyllum*
Genus: 60 species of cool or temperate regions; 6 in Atlantic Canada
Family: Rosaceae (Rose)
Etymology: A name used by Pliny; large-leaved
Habitat: Usually in shaded wet ground such as intervales, along streams, or in damp woods
Range: Labrador to Ontario, south to New York; Alaska
Longevity: Perennial
Flowering: July–August

Features: The five-petalled yellow flowers have many stamens. The terminal segment of the basal leaves is much larger than the lateral lobes, almost round and heart-shaped at the base, while the leaves on the lower stem are three-parted.
Fruit: Numerous achenes
Height: 3–100 cm.
Horticulture: Moist, sandy soil (pH 6.0–7.0); full sun
Similar Species: *G. rivale* has greenish- or purplish-cream-coloured petals.

Bakeapple
also called cloudberry

Latin Name: *Rubus Chamaemorus*
Genus: 120 species in North America, 16 of them in Atlantic Canada
Family: Rosaceae (Rose)
Etymology: Latin *ruber*, red; old generic name for this species
Habitat: Sphagnum bogs, barrens, meadows near the coast, headlands
Range: Greenland to Alaska, south to Maine and New Hampshire; Eurasia
Longevity: Perennial
Flowering: July
Features: White solitary flower is more than 1 cm. wide and has 5 petals and many pistils and stamens. There are 1–3 leaves, which have 3–5 rounded lobes.
Fruit: An aggregation of drupelets
Height: 15–20 cm.
Ecology: The berries are eaten by birds and chipmunks.
Uses: The raw berries are at their best when not quite ripe and eaten with sugar and cream. When fully ripe, they make a good jam.
Note: There are separate male and female plants.

References: Kaurin, A., C. Stushnoff, and O. Junttila, 1982. Vegetative growth and frost hardiness of cloudberry (*Rubus chamaemorus*) as affected by temperature and photoperiod. *Physiologia Plantarum*, vol. 55(2), 76–81.

Warr, H.J. and D.R. Savory, 1979. Germination studies of bakeapple (cloudberry) (*Rubus chamaemorus*) seeds. *Canadian Journal of Plant Science*, vol. 59(1), 69–74.

Flowering raspberry
also called thimbleberry

Latin Name: *Rubus odoratus*
Genus: About 120 species in North America; 16 in Atlantic Canada
Family: Rosaceae (Rose)
Etymology: The Roman name, *ruber*, for red; fragrant
Habitat: Edges of woods, ravines, and wild hedgerows
Range: Nova Scotia to southern Ontario, south to Georgia and Tennessee
Longevity: Perennial, shrub
Flowering: Late June – August
Features: Large showy flowers are rose–purple, with 5 rounded petals and numerous stamens and pistils. Leaves have 3–5 lobes, while the branches have glandular hairs.
Fruit: Drupelets
Height: To 15 cm.
Ecology: Some birds eat the berries.
Uses: Fruit is edible but dry, acidic, and insipid.
Similar Species: *R. Chamaemorus* has white flowers.

References: Fassett, N.C., 1941. Moss collections: *Rubus odoratus* and *R. parviflorus. Annals of the Missouri Botanical Garden*, vol. 28, 299–374.

Blackberry

Latin Name: *Rubus* spp.
Genus: About 400 kinds of blackberries have been named; uncertain number in Atlantic Canada, but at least 10 are identified
Family: Rosaceae (Rose)
Etymology: From the Roman name *ruber*, red
Habitat: Old fields, pastures, softwood cutovers, fence rows, and edges of woodlands
Range: Newfoundland to Minnesota, south to Pennsylvania
Longevity: Perennial
Flowering: June–August
Features: White five-petalled flowers have numerous stamens and pistils and occur in a panicle. The raspberry-like leaves have 3–5 leaflets and are attached to sprawling, thorny stems.
Fruit: Black drupelets
Height: 180 cm.

Uses: Edible, like raspberries. The high tannin content makes the roots useful as an astringent; a tea can be made from them to stop secretions. The berries can be used as a fruit juice to control diarrhoea or in making wine.
Note: Like raspberries, blackberries hybridize easily and are a most difficult group to handle taxonomically.
Similar Species: Raspberries, *Rubus* spp.

References: Federer, C.A., 1977. Leaf resistance and xylem potential differ among broadleaved species. *Forest Science*, vol. 23(4), 411–19.

Parisio, S., 1982. Uptake by plants of calcium, magnesium and nickel on a serpentine soil in Staten Island, New York (*Andropogon scoparius, Lonerica japonica, Rubus alleghe-niensis*). Proceedings Staten Island Institute of Arts and Sciences, vol. 31(2/3), 70–6.

Canadian burnet

Latin Name: *Sanguisorba canadensis*
Genus: About 20 species of the north temperate zone; 3 in Atlantic Canada
Family: Rosaceae (Rose)
Etymology: Lat. *sanguis*, blood, and *sorbere*, to absorb (from use in folk medicine); Canadian
Habitat: Wet meadows, bogs, and well-drained swamps
Range: Labrador and Newfoundland to Michigan and locally south to Long Island, Delaware, and the mountains of Georgia
Longevity: Perennial
Flowering: July–September

Features: Small white flowers have four-petaloid sepals and occur in dense spikes at the top of tall stems. The leaves are pinnately compound and toothed.
Fruit: Solitary achene
Height: 50–150 cm
Horticulture: Moist, cool, acidic soil; partial shade
Note: The lower flowers open first, leaving the upper part of the spike in bud.
Similar Species: *S. officinalis* and *S. minor* occur only in localized areas of Nova Scotia.

White rose

Latin Name: *Rosa multiflora*
Genus: More than 100 species have been described in the northern hemisphere; 10 in Atlantic Canada.
Family: Rosaceae (Rose)
Etymology: The ancient Latin name; many-flowered
Habitat: Clearings, roadsides, borders of woods; often planted in hedges
Range: Introduced and naturalized from eastern Asia
Longevity: Perennial
Flowering: June
Features: A tall, arching shrub with numerous white five-petalled flowers that have many stamens. Leaves consist of 7–9 leaflets and are on a usually prickly stem.
Fruit: Achene
Height: To 1.5 cm
Uses: Rose-hips can be used in herbal tea, candied, or used for jelly.
Note: Occasionally planted for hedges

References: Davies, F.T. Jr., Y. Fann, J.E. Lazarte, and D.R. Paterson, 1980. Bench chip budding of field roses (*Rosa hybrida, Rosa multiflora*). *HortScience*, vol. 15(6), 817–18.

Fawcett, R.S., 1980. Multiflora rose (*Rosa multiflora*, cultural and chemical control). *Weeds Today*, vol. 11(1), 22–3.

Common wild rose

Latin Name: Rosa virginiana
Genus: More than 100 species are native to the northern hemisphere; 9 in Atlantic Canada.
Family: Rosaceae (Rose)
Etymology: The ancient Latin name; Virginian
Habitat: Occurs in wet pastures, thickets, dike lands, swamps, and head of salt marshes
Range: Newfoundland to southern Ontario, south to Alabama
Longevity: Perennial
Flowering: July–August
Features: The pink flowers have 5 petals and many stamens. They occur singly or in corymbs, always borne on branches from the old wood. Flattened prickles are sometimes present on the stout, much-branched stem. Five to 9 leaflets occur on the alternate leaves and are without pubescence.

Fruit: Achenes in red rose-hip.
Height: Up to 2 m.
Ecology: The fleshy fruit provides winter food for many birds and mammals. Rose-hips are a favourite food for pheasants, grouse, and robins.
Uses: Petals can be used for jam, rose-hips for jelly, and the rose-hips and flowers for herbal tea. They have a high content of vitamin C; highly useful as an emergency food.
Horticulture: Easily transplanted as it spreads by suckers
Similar Species: R. carolina (with which it hybridzes) is more slender and rarely more than 1 m. high.

References: Thomas, G.S., 1977. The 'Rose d'Amour' and the 'D'orsay Rose': hybrids of Rosa virginiana. Rose Annual Royal National Rose Society, 27-8.

Pin cherry
also called bird cherry

Latin Name: Prunus pensylvanica
Genus: About 200 species, chiefly of the north temperate zone; 9 in Atlantic Canada
Family: Rosaceae (Rose)
Etymology: The ancient Latin name of plum; Pennsylvanian
Habitat: Edges of fields, ditches, barrens, thickets, and burnt-over land
Range: Labrador to British Columbia, south to North Carolina
Longevity: Perennial (shrub)
Flowering: May 25 – June 20
Features: Several to many white five-petalled flowers occur with many stamens. The oblong-lanceolate leaves have sharp, in-turned teeth.
Fruit: Red, fleshy drupe with a large stone

Height: Up to 12 m.
Ecology: Cherries are eaten by birds.
Uses: The fruit may be eaten fresh but also makes excellent jelly, preserves, juice, syrup, and wine. It is delicious when used for pie, ice cream, and baked products.
Note: The twigs and leaves of all cherries can be fatal, containing a compound that releases cyanide
Similar Species: P. virginiana flowers are in a raceme; P. Avium occasionally grows as an escape in the Annapolis Valley.

References: Federer, C.A., 1977. Leaf resistance and xylem potential differ among broadleaved species. *Forest Science,* vol. 23(4), 411–19.

Garden-lupin

Latin Name: *Lupinus polyphyllus*
Genus: About 90 species in western North America; 2 in Atlantic Canada
Family: Fabaceae (Pea)
Etymology: Ancient name from *lupus*, wolf; many-leaved
Habitat: Grown as an ornamental, but has escaped into ditches and fields
Range: British Columbia; introduced, from Prince Edward Island and Nova Scotia to New England
Longevity: Perennial
Flowering: June 15 – July
Features: Showy pea-like flowers are numerous in a terminal raceme or spike. They are usually purple but can also be white or pink. The leaves are divided into 10–17, palmately compound, elongate leaflets, which are glabrous to slightly hairy beneath.
Fruit: Oblong legume
Height: Up to 120 cm.
Horticulture: Well-drained, moderately fertile soil (pH 7.0); full sun
Note: Another species, bluebonnet, is the state flower of Texas. Some species are reported to be poisonous to sheep.
Similar Species: *L. nootkatensis* has fewer leaves, which are hairy underneath, and is localized in Yarmouth area.

Red clover

Latin Name: *Trifolium pratense*
Genus: About 300 species in temperate northern hemisphere; 7 in Atlantic Canada
Family: Fabaceae (Pea family)
Etymology: Latin *tres*, three, *folium* leaf; of meadows
Habitat: Fields and meadows; rarely persisting except along roadsides and dwellings
Range: Widely naturalized in temperate North America; introduced from Europe
Longevity: Biennial or short-lived perennial
Flowering: Early summer
Features: Reddish flowers are sessile in large spherical heads. The leaves are three-parted, with a chevron pattern on each leaflet.
Fruit: Legume
Height: 5–40 cm.
Ecology: Seeds and leaves are eaten by many birds and mammals.
Uses: The entire plant is edible. Leaves can be used in a salad or as a pot-herb and are high in protein. Flowers are popular in medicinal tea. The seeds and dried flowers have been used as a bread food during famine in the British Isles.
Note: State flower of Vermont. Clover produces nitrogen and enriches the soil.
Similar Species: Var. *sativum* is cultivated in hayfields. *T. repens* has small, white flower clusters, while *T. hybridum* has pink and white flowers.

Low hop-clover

Latin Name: *Trifolium procumbens*
Genus: About 300 species in temperate northern hemisphere; 7 in Atlantic Canada
Family: Fabaceae (Pea)
Etymology: Latin *tres*, three; *folium*, leaf; trailing
Habitat: Along roadsides, in towns and waste places
Range: Nova Scotia and Prince Edward Island to North Dakota, south to Georgia; Pacific slope; naturalized from Europe
Longevity: Annual
Flowering: July–September
Features: The dense heads have 20–30 yellow flowers that turn brown with age. Three leaflets occur on depressed stems or wide-spreading branches.
Fruit: Legume
Height: Up to 30 cm.
Uses: The plant is good to eat fresh after dipping in salted water. It can also be cooked as a vegetable or used in a soup, custard, salad, or lemonade.
Note: It is reputed to have been recognized in Ireland as the true shamrock for many years.
Similar Species: *T. agrarium* has a sessile terminal leaflet, *T. dubium* is a much smaller plant, and *T. arvense* has grey, silky-plumose flowers.

Bird's-foot-trefoil

Latin Name: Lotus corniculatus
Genus: 140 species of Eurasia and western North America; 2 in Atlantic Canada
Family: Fabaceae (Pea)
Etymology: The ancient Greek name restricted to these plants by Linnaeus; horned
Habitat: Planted as forage, but occasionally seen persisting along roadsides and in fields
Range: Newfoundland to British Columbia, south to Virginia
Longevity: Perennial
Flowering: July–September
Features: The inflorescence has fewer than 8 yellow pea-like flowers on long erect peduncles. Five leaflets, two of which are stipules, occur on the pinnately compound leaflets.
Fruit: Pod
Height: To 50 cm.
Note: An introduced field plant
Similar Species: L. uliginosus has well-developed runners and 7–11 flowers.

References: DeGrandi-Hoffman, G. and C.H. Collison, 1982. Flowering and nectar secretion as they relate to honeybee foraging activity in birdsfoot trefoil (Lotus corniculatus) (Apis mellifera, ecological and physiological aspects, Pennsylvania). Journal of Agricultural Research, vol. 21(4), 199–207.

Hankins, B.J., C.L. Rhykerd, G.O. Mott, and B.O. Blair, 1980. History of (Lotus corniculatus L.) in Indiana (Plant ecology). Proceedings Indiana Academy of Science, vol. 89, 151–3.

Tufted vetch

Latin Name: *Vicia Cracca*
Genus: About 200 widely distributed species; 7 in Atlantic Canada
Family: Fabaceae (Pea)
Etymology: The classical Latin name; Italian name, *Cracca*, applied by Rivinius
Habitat: Meadows, pastures, gardens, railroads, waste places, grain fields, and row crops
Range: In all provinces; introduced from Europe
Longevity: Perennial
Flowering: Late June – August
Features: The blue-purple butterfly-like flowers grow in close, many-flowered, one-sided spikes. Leaves have 20–24 leaflets and a tendril for climbing.
Fruit: Legume
Height: Up to 2 m.
Uses: All vetches make good hay and enrich soil by building up nitrates in the roots. Useful as an emergency food.
Similar Species: *V. villosa* is occasionally sown and persists for a while in open sandy soil.

References: Klebesadel, L.J., 1980. Birdwatch (*Vicia cracca*, legume)—forage crop, ground cover, ornamental, or weed? (in Alaska). *Agroborealis*, Alaska, vol. 12(1), 46-9.

Beach pea

Latin Name: Lathyrus japonicus
Genus: 150 species in northern hemisphere and South America; 4 in Atlantic Canada
Family: Fabaceae (Pea)
Etymology: Latin *la*, very, and *thuros*, passionate (the original plant was reported to be an aphrodisiac); Japanese
Habitat: Sand dunes and upper beaches, but may invade elsewhere
Range: Newfoundland, south along the coast to New Jersey; inland around the Great Lakes; British Columbia to California
Longevity: Perennial
Flowering: Early summer
Features: Purple butterfly-shaped flowers occur in a cluster. The leaves are divided into 3–5 pairs of thick oblong leaflets with a tendril.
Fruit: Legume
Height: 30 cm.

Uses: The young peas can be used in salads but are dry with a slightly disagreeable taste. However, they are useful as an emergency food. The young shoots can be boiled and made into a salad.
Note: The peas are supposed to contain a poisonous alkaloid, and the herbage is probably poisonous, as are others in this genus.
Similar Species: Var. *glaber* is glabrous, and var. *pellitus* is pubescent

References: Beesley, S., 1979. The sea pea, *Lathyrus japonicus* Willd. in County Kerry. *The Irish Naturalists' Journal*, vol. 19(9), 328.

Lechowicz, M.J., L.E. Hellens, and J.P. Simon, 1980. *Lathyrus japonicus*. *Canadian Journal of Botany*, vol. 58(14), 1522–4.

McMillan, N.F., 1981. The habitat of the sea pea, *Lathyrus japonicus* Willd. (British Isles, Canada). *The Irish Naturalists' Journal*, vol. 20(5), 206–7.

Wild pea

Latin Name: *Lathyrus palustris*
Genus: 150 species in northern hemisphere and temperate South America and Africa; 4 species in Atlantic Canada
Family: Fabaceae (Pea)
Etymology: Reputed to be from the Latin *la*, very, and *thuros*, passionate (the original plant was supposed to have aphrodisiac properties); of marshes
Habitat: Wet places, such as damp thickets or edges of marshes near shores
Range: Newfoundland to Manitoba, south to New York and Missouri; Oregon and northwards to British Columbia.
Longevity: Perennial
Flowering: June 20 – July

Features: The 2-9 pea-like flowers have purple corollas. Leaflets occur in 2-6 pairs, are linear or narrowly oblong, and have stipules with 1 basal lobe.
Fruit: Legume
Height: 30-90 cm.
Uses: The tender young seeds can be used like a green pea but are dry and disagreeable in taste.
Note: The herbage of many *Lathyrus* is poisonous; one should be careful to test before eating.
Similar Species: *L. japonicus* has oval leaflets; several varieties of *L. palustris* have been described.

References: Vaughan, I.M. and J. Watsonia, 1972. *Lathyrus palustris* L.—in Carmarthenshire. *Proceedings Botanical Society of the British Isles*, vol. 9(2), 137.

Wood sorrel

Latin Name: *Oxalis montana*
Genus: 800 species, most of which are in South Africa and South America; 4 are in Atlantic Canada
Family: Oxalidaceae (Wood sorrel)
Etymology: Greek *oxus*, sour (referring to acidic taste); of the mountains
Habitat: Damp woods, mossy banks, along ravines, and in wooded swamps
Range: Southern Newfoundland to Ontario and Minnesota, south to North Carolina and Tennessee
Longevity: Perennial
Flowering: Early June – July
Features: The one flower per stalk has 5 white petals veined with rose or purple. Later flowers are cleistogamous. Its leaves are all basal, shamrock-shaped, and are green above and purplish beneath.
Fruit: Elongated capsule
Height: 15 cm.
Ecology: The seeds are eaten by some birds, while hares and deer eat the leaves.
Uses: The leaves have a pleasant acidic taste and can be used in salads, as a pot-herb, or as a lemonade-like drink when steeped in water.
Horticulture: Grows well in acidic soil (pH 4.0–5.0); partial shade
Note: The plant tastes like sheep sorrel (*Rumex Acetosella*) but is not related.
Similiar Species: *O. stricta* is taller, with 1–3 yellow flowers; also closely related to the Eurasian *O. Acetosella*

Yellow wood sorrel

Latin Name: *Oxalis stricta*
Genus: 800 species of which 100 occur in North America; 4 in Atlantic Canada
Family: Oxalidaceae (Wood sorrel)
Etymology: Greek *oxus*, sour (referring to acidic taste); erect
Habitat: Roadsides, thickets, waste ground, fields, and near dwellings
Range: Prince Edward Island and Nova Scotia to Saskatchewan, south to Virginia and Arizona; introduced in Europe
Longevity: Perennial
Flowering: June–July
Features: One to 3 small yellow flowers occur per stalk on short branching pedicels. Leaves are divided into 3 clover-like leaflets, and stems are erect or reclining
Fruit: Elongated capsule
Height: Up to 50 cm.
Uses: Pleasant acidic leaves can be used in salads, as a pot-herb, or steeped in water to form a lemonade-like drink.
Note: Numerous forms have been described based on the pubescence of the stems, leaves, and pedicels.
Similar Species: *O. montana* has 1 pink-striped white flower.

Spotted-touch-me-not
also called spotted jewelweed

Latin Name: *Impatiens capensis*
Genus: About 400 species, mostly in India; 4 species in Atlantic Canada
Family: Balsaminaceae (Touch-me-not)
Etymology: Latin *impatiens*, impatient (refers to sudden bursting of capsules when touched); of the Cape (origin was wrongly thought to be the Cape of Good Hope)
Habitat: Moist places, such as along brooks, ditches, and thickets with high organic matter and nitrogen
Range: Newfoundland to Alaska, south to Florida and Oklahoma
Longevity: Annual
Flowering: July–August
Features: Orange flowers have reddish-brown spots, a strongly incurved spur, and occur in axillary racemes. Plant is much branched with alternate leaves and has watery juice.
Fruit: Capsule, which, when it bursts, projects seeds for several metres
Height: Up to 150 cm.
Ecology: Nectar provides food for bees and humming birds. Seeds are eaten by mice and some birds and leaves and stems by hares.
Uses: There is confusion as to edibility, but spring shoots are often eaten after boiling in two waters. Reported as a source of anti-fungal extracts. Stem juice is soothing for nettle stings and poison ivy rash.
Horticulture: Moist, rich, semi-shaded soil
Note: The plant has the reputation of being emetic and poisonous to livestock.
Similar Species: Numerous colour forms have been described, and varieties are common. *I. pallida* has pale yellow flowers but is not common.

References: Howell, N., 1981. The effect of seed size and relative emergence time on fitness in a natural population of *Impatiens capensis* Meerb. (Balsaminaceae). *American Midland Naturalist*, vol. 105(2), 312–20.

Musk-mallow

Latin Name: Malva moschata
Genus: About 30 species, chiefly in Eurasia; 6 in Atlantic Canada
Family: Malvaceae (Mallow)
Etymology: Greek *malache* or *moloche*, indicating the emollient leaves; musky
Habitat: Localized in waste places, old hay fields, along roadsides, and in old gardens
Range: Newfoundland to Manitoba, south to Virginia; British Columbia; introduced from Europe
Longevity: Perennial
Flowering: Late June – July
Features: The pink or white five-petalled flowers are clustered at the stem summit and have numerous stamens united in a column. The stem leaves are five-parted, with the divisions cleft into linear lobes.
Fruit: Capsule
Height: 20–70 cm.
Uses: May be eaten as a pot-herb. The mucilaginous juice thickens soup, and the scalloped fruits are edible.
Note: This plant has wandered from garden to roadside.

References: Kearney, T.H., 1951. The American genera of Malvaceae. *American Midland Naturalist*, vol. 46, 93–131.

Common St. John's wort

Latin Name: *Hypericum perforatum*
Genus: Nearly 300 species around the world; 9 in Atlantic Canada
Family: Hypericaceae (St. John's wort)
Etymology: Greek *hypericon*, ancient name of another plant; perforated (alluding to spotted petals)
Habitat: Light sandy soils, fields, gravelly river edges, roadsides
Range: Newfoundland to British Columbia and south into the United States; naturalized from Europe
Longevity: Perennial
Flowering: July 10 – August
Features: Bright yellow flowers have 5 petals with dark dots on the edges and numerous stamens. The opposite, untoothed leaves have glands and are linear-oblong
Fruit: Many-seeded capsule
Height: 40–80 cm.

Ecology: Used as food by ducks and ruffed grouse
Uses: Superstitious people used to gather this plant on St. John's Eve, hanging it at the doors and windows to ward off evil spirits and thunder. In early times an ointment was made from its flowers. It was also reported to be an effective remedy for melancholia.
Note: This plant contains a poison that affects white-haired animals when they are exposed to the sun after ingesting it.
Similar Species: *H. punctatum* has a compact inflorescence and elliptical leaves.

References: Gillett, J.M. and N.K.B. Robson, 1981. The St. John's-wort of Canada (*Guttiferae*). *National Museums of Natural Sciences Publications in Botany*, no. 2.

Mabbott, D.C., 1920. Food habits of seven species of shoal water ducks. *U.S. Department of Agriculture Bulletin #862.*

Blue violet

Latin Name: *Viola cucullata*
Genus: 500 species, nearly cosmopolitan; 17 in Atlantic Canada
Family: Violaceae (Violet)
Etymology: Latin *Viola*, the classical name; hooded (referring to in-rolled young leaves.)
Habitat: Wet fields, swamps, rocky beaches, and meadows
Range: Newfoundland to British Columbia, south to Georgia and Tennessee
Longevity: Perennial
Flowering: May–July
Features: The plant is stemless, with the leaves and flower stocks growing directly from rootstocks. Purplish-blue flowers occur on peduncles longer than the heart-shaped leaves and have a spurred petal.
Fruit: Ellipsoid capsule
Height: 1–10 cm.
Uses: The leaves are a nice addition to a salad or can be cooked as greens. Flowers have been used in jams and syrups and are supposed to have properties to soothe the digestive tract and suppress a cough. Flowers and leaves are high in vitamins C and A.
Horticulture: Rich humus soil; partial shade
Note: Floral emblem of New Brunswick. Seeds are produced in cleistogamous flowers.
Similar Species: Forma *prionosepala* has hairier leaves; var. *microtitis* is rare. *V. nephrophylla* has flowers on peduncles equal to or shorter than the leaves. *V. septentrionalis* and *V. firmbriatula* are hairy plants.

References: Russell, N.H., 1965. Violets (*Viola*) of central and eastern United States: an introductory survey. *Sida*, no. 2, 1–113.

Small white violet

Latin Name: *Viola pallens*
Genus: About 500 species (nearly cosmopolitan); 17 in Atlantic Canada
Family: Violaceae (Violet)
Etymology: Latin *Viola*, the classical name; pale
Habitat: Moist ground in meadows, bogs, borders of lakes, and thickets
Range: Labrador to Alaska, south to South Carolina and Colorado
Longevity: Perennial
Flowering: May – early June
Features: The small white flowers have purple veins and occur on peduncles longer than the leaves. Leaves are heart-shaped and glabrous on both sides.
Fruit: Ellipsoid capsule
Height: 1–8 cm.
Uses: Often used as a spring tonic since the leaves and blossoms are high in vitamin C. Vitamin A occurs in the leaves.
Horticulture: Rich humus soil; partial shade
Note: The small white violet is one of the first signs of spring. Unlike many wildflowers, picking the blossoms does not harm it, for the blossoms seldom produce seed. The seeds are produced by the cleistogamous flowers.
Similiar Species: *V. incognita* has pubescent leaves and petioles.

References: Newell, S.J., O.T. Solbrig, and D.T. Kincaid, 1981. Studies on the population biology of the genus *Viola*, III. The demography of *Viola blanda* and *Viola pallens*. The *Journal of Ecology*, vol. 69(3), 997–1016.

Schellner, R.A., S.J. Newell, and O.T. Solbrig, 1982. Studies on the population biology of the genus *Viola*, IV. Spatial pattern of ramets and seedlings in three stoloniferous species (*Viola blanda, Viola pallens, Viola incognita*). *The Journal of Ecology*, vol. 70(1), 273–90.

Purple loosestrife

Latin Name: *Lythrum Salicaria*
Genus: About 30 species, mostly in the north temperate zone; 1 in Atlantic Canada
Family: Lythraceae (Loosestrife)
Etymology: From *lytron*, a name used by Dioscorides; old generic name
Habitat: Wet meadows, shores, and upper edges of salt marshes
Range: Nova Scotia to Ontario, south to New York, Indiana, and Missouri; naturalized from Europe
Longevity: Perennial
Flowering: July–August
Features: Purple flowers are numerous in long spikes and are rather showy. The leaves are opposite and somewhat heart-shaped at the base.
Fruit: Many-seeded pod
Height: Up to 1 m.
Note: This plant has become much more widespread in recent years and is classed as an aggressive weed. The stamens and pistils occur in three different lengths, making it possible for pollen from any given set of stamens to fertilize a pistil of corresponding length.
Similar Species: Var. *tomentosum* has a softly white tomentose calyx and bracts

Fireweed
also called large willow-herb

Latin Name: Epilobium angustifolium
Genus: 100 species, in the boreal and temperate northern hemisphere, with 11 present in Atlantic Canada
Family: Onagraceae (Evening Primrose)
Etymology: Greek *epi*, upon, and *lobon*, pod; narrow-leaved
Habitat: Burnt-over areas, along fence rows, edges of thickets, and in waste places
Range: Greenland to Alaska, south to North Carolina and California; Eurasia
Longevity: Perennial
Flowering: July 10 – August
Features: The pinkish four-petalled flowers occur in an elongate raceme on which the bottom flowers blossom first. Leaves are long and glabrous.
Fruit: Long, slender capsule
Height: 1 m. or higher
Uses: The young leaves and stems may be eaten as greens, and the older ones may be made into tea. The roots are reported to have been cooked by the Indians to make a cure for boils.
Horticulture: Moist, rich humus soil (pH 5.0–6.0); light shade
Note: Fireweed is the floral emblem of the Yukon. It spreads both by runners and seed dispersal.
Similar Species: Forma *albiflorum* has white flowers; forma *spectabile* has white petals with red sepals.

References: Addicott, J.F., 1978. The population dynamics of aphids on fireweed (*Epilobium angustifolium*): a comparison of local populations and metapopulation. *Canadian Journal of Zoology*, vol. 56(12), 2554–64.

Henderson, G., P.G. Holland, and G.L. Werren, 1979. The natural history of a subarctic adventive: *Epilobium angustifolium* L. *(Onagraceae)* at Schefferville, Quebec. *Le naturaliste canadien*, vol. 106(4), 425–37.

Evening primrose

Latin Name: *Oenothera biennis*
Genus: Probably 150 species of North and South America; 6 in Atlantic Canada
Family: Onagraceae (Evening primrose)
Habitat: Pastures, shoulders of roads, and in waste places
Range: Newfoundland to British Columbia, south to Florida and Tennessee
Longevity: Biennial
Flowering: June–September
Features: The showy yellow four-petalled flowers are at their best in the evening and fade by the morning. The plant has alternate leaves and forms winter rosettes in its first year.
Fruit: Many-seeded capsule
Height: 15 cm.–150 cm.
Uses: If picked not too early in autumn or too late in the spring and if cooked in two waters, the root is edible as a vegetable.

Note: Evening primrose is self-pollinating, but occasionally cross-pollination occurs; consequently, a number of races are found.
Similiar Species: Three forms have been described in Atlantic Canada, as well as two varieties and many races.

References: Dickey, J.L. and M. Levy, 1979. Development of powdery mildew (*Erysiphe polygoni*) on susceptible and resistant races of *Oenothera biennis*. *American Journal of Botany*, vol. 66(9), 1114–17.

Gross, K.K. and P.A. Werner, 1982. Colonizing abilities of "biennial" plant species in relation to ground cover: implications for their distributions in a successional sere (*Verbascum thapsus*, *Oenothera biennis*, *Daucus carota*, *Tragopogon dubius*, Michigan). *Ecology*, vol. 63(4), 921–31.

Sundrops

Latin Name: *Oenothera perennis*
Genus: Probably 150 species in North and South America; 6 in Atlantic Canada
Family: Onagraceae (Evening primrose)
Etymology: Name used by Theophrastus for another plant; perennial
Habitat: Roadsides, meadows, and light sandy soils
Range: Newfoundland to Manitoba, south to Delaware, Georgia, and Ohio
Longevity: Perennial
Flowering: July–September

Features: The yellow four-petalled flowers often have a reddish calyx. There is a basal rosette of spatulate leaves, while the stem leaves are linear-lanceolate.
Fruit: Winged capsule
Height: 10–50 cm.
Horticulture: Thin, well-drained soil (pH 6.0–7.0); full sun
Note: A day bloomer
Similar Species: *O. tetragona*, which is not common

Bristly sarsaparilla

Latin Name: Aralia hispida
Genus: About 30 species in North America and Asia; 3 in Atlantic Canada
Family: Araliaceae (Ginseng)
Etymology: Name from French Canadian *Aralie*, given by Québec physician Sarrasin; with straight hairs
Habitat: Dry woodlands, old forests, and rocky, sandy clearings
Range: Newfoundland to British Columbia, south to Georgia, Illinois, and Colorado
Longevity: Perennial
Flowering: May 25 – June

Features: Small white-to-greenish flowers are polygamous, five-petalled, and occur in several umbels on a peduncle. The stem is bristly at the base, and the leaves are doubly compound.
Fruit: Drupe
Height: 40–80 cm.
Note: The fruit is aromatic or spicy.
Similar Species: *A. nudicaulis* has stem almost absent; *A. racemosa* is much smaller.

References: Smith, A.C., 1936. Notes on North American Araliaceae. *Brittonia*, vol. 2, 247–61.

Wild sarsaparilla

Latin Name: Aralia nudicaulis
Genus: About 30 species in North America and Asia; 3 in Atlantic Canada
Family: Araliaceae (Ginseng)
Etymology: Name from the French Canadian *Aralie*; naked stem
Habitat: Dry woodlands and old forests
Range: Newfoundland to British Columbia, south to Georgia, Illinois, and Colorado
Longevity: Perennial
Flowering: May 25 – June
Features: The insignificant white-to-greenish flowers occur in 3 umbels on a naked scape, hidden under the foliage. Stem is almost absent, but has leaves with 3 main divisions, which are further divided.
Fruit: Blue-black berry
Height: 20–40 cm.
Ecology: The berries are eaten by some birds and mammals.

Uses: Berries are used to make wine and are edible when cooked. Roots are used in herbal tea, root beer, as an emergency food, and as a substitue for the official sarsaparilla. In pioneer days, a solution from the root was applied to legs of exhausted horses.
Note: One of the most common flowering plants in early summer
Similar Species: A. hispida occurs in open areas and has a tall stem with bristles at the bottom.

References: Barret, S.C.H. and K. Helenurm, 1981. Floral sex ratios and life history in *Aralia nudicaulis (Araliaceae)* (wild sarsaparilla). *Evolution*, vol. 35(4), 752–62.

Barrett, S.C.H. and J.D. Thomson, 1982. Spatial pattern, floral sex ratios, and fecundity in dioecious *Aralia nudicaulis (Araliaceae)*. *Canadian Journal of Botany*, vol. 60(9), 1662–70.

Bawa, K.S., C.R. Keegan, and R.H. Voss, 1982. Sexual dimorphism in *Aralia nudicaulis* L. *(Araliaceae)*. *Evolution*, vol. 36(2), 371–8.

Cow parsnip

Latin Name: *Heracleum lanatum*
Genus: About 60 species of the north temperate zone; 2 in Atlantic Canada
Family: Apiaceae (Parsley)
Etymology: Dedicated to Hercules; woolly
Habitat: Intervales, wet meadows, or alluvial soil by brooks
Range: Labrador to Alaska, south to New England, and in the mountains to Georgia and Ohio
Longevity: Perennial
Flowering: Early July
Features: White five-petalled flowers occur in a large flat umbel. The large leaves are irregularly cut and divided into 3 main sections, which are woolly beneath.
Fruit: Mericarp
Height: To 2.8 m.

Ecology: Seeds eaten by many birds
Uses: Before the plant flowers, the young leaf stalks and stems can be stewed like celery. The large root can be cooked and eaten as a vegetable. In California, Indians were reported to have used the hollow basal portion of the plant as a salt substitute.
Horticulture: Moist rich soil; sun or partial shade
Note: It is shunned by cattle because of its bitter juice and strong odour.

References: Nakata, H., Y. Sashida, and H. Shimomura, 1982. A new phenolic compound from *Heracleum lanatum* Michx. var. *nippinicum* Hara. *Chemical and Pharmaceutical Bulletin*, vol. 30(12), 4554–6.

Shimomura, H., Y. Sashida, H. Nakata, J. Kawasaki, and Y. Sto, 1982. Plant growth regulators from *Heracleum lanatum*. *Phytochemistry*, vol. 21(9), 2213–15.

Queen Anne's lace
also called wild carrot

Latin Name: Daucus carota
Genus: 60 species, global in distribution; 1 in Atlantic Canada
Family: Apiaceae (Parsley)
Etymology: Greek *daukos*, the ancient Greek name; old generic name for carrot
Habitat: Common in hay fields, along roadsides, and in waste places
Range: Throughout North America, naturalized from Europe
Longevity: Biennial
Flowering: July–September
Features: The five-petalled flowers occur in flat-topped umbels that often have a purple centre flower. The leaves are pinnately compound and finely dissected. Both the leaves and stem are hairy.
Fruit: Schizocarp
Height: 50–100 cm.
Ecology: The seeds are eaten by some birds and mice.
Uses: The first-year roots are reported to be edible. However, the roots are said to be somewhat poisonous; care should be taken not to confuse it with another plant, Caraway.
Note: Queen Anne's lace forms fertile hybrids when crossed with the cultivated carrot and is believed to be the stock from which the garden carrot was raised.
Similar Species: Carum carvi has crowded leaflets, but leaves and stem are not hairy.

References: Dale, H.M., 1974. The biology of Canadian weeds, 5: *Daucus carota. Canadian Journal of Plant Science,* vol. 54, 673–85.

Bunchberry
also called pigeon berry or cracker berry

Latin Name: *Cornus canadensis*
Genus: 40 species; 5 occur in Atlantic Canada
Family: Cornaceae (Dogwood)
Etymology: Latin *cornus*, a horn (alluding to a hard-wooded European species); Canadian
Habitat: Heaths, barrens, mature bogs, thicket edges, and is a woodland pioneer
Range: Labrador to Alaska, south to North Carolina, West Virginia, and California
Longevity: Perennial
Flowering: June
Features: The small, numerous, greenish-white flowers are subtended by 4 white petaloid bracts and are in a solitary, flat-topped inflorescence. This low herb has its leaves whorled near the top of the stem, except for a distinctive pair of small leaves at mid-stem.
Fruit: Red drupe
Height: 10–20 cm.

Ecology: Berries are used as food by chipmunks and mice and are much loved by birds.
Uses: Berries are edible and rather tasteless, but are good when added to pudding or mixed with other fruit. They were once popular for fattening livestock.
Horticulture: Cool, acidic soil (pH 4.0–5.0); filtered light
Similiar Species: A number of forma have been described; *racemosa, intraverticillata, medioloides, elongata,* and *purpurascens.*

References: Bain, J.F. and K.E. Denford, 1979. The flavonoid glycosides of *Cornus canadensis* L. and its allies in northwestern North America. *Experientia*, vol. 35(7), 863-4.

Hall, I.V. and J.D. Sibley, 1976. The biology of Canadian weeds, 20: *Cornus canadensis* L. *Canadian Journal of Plant Science*, vol. 56, 885–92.

One-flowered shinleaf

Latin Name: *Moneses uniflora*
Genus: A single boreal species that occurs in Atlantic Canada
Family: Pyrolaceae (Wintergreen)
Etymology: Greek *monos*, single, and *hesis*, delight, referring to the showy flower; one-flowered
Habitat: In deciduous and mixed woodlands, sometimes in coniferous woods
Range: Labrador to Alaska, south to Pennsylvania and Minnesota; Eurasia

Longevity: Perennial
Flowering: June 20 – July 20
Features: The solitary waxy-white flower is on a stalk that nods in flower but straightens up in fruit. Ten stamens lie against the 5 petals. The leaves are mostly basal and almost round.
Fruit: Capusle
Height: 3–13 cm.
Note: The smallest of the wintergreens

Indian-pipe
also called corpse-plant

Latin Name: *Monotropa uniflora*
Genus: 2 species, a small family of the northern hemisphere; both species are in Atlantic Canada
Family: Ericaceae (Heath)
Etymology: Greek *monos*, one, and *tropos*, turn (refers to the bent over flower head); one-flowered
Habitat: Often found in dense shade in climax or old coniferous forest and is also common in leaf mould in mixed or deciduous woods
Range: Labrador to British Columbia, south to Florida and Mexico; Asia
Longevity: Perennial
Flowering: July–August
Features: Flower heads are solitary and bent over, but erect when in seed. Plant is white and smooth and turns black in dying. Colourless bracts are on the stem instead of leaves. The plant has corolla of 5 petals and 10 stamens with yellow anthers.
Fruit: Brown many-seeded capsule
Height: 5–30 cm.
Uses: It can be eaten as a cooked vegetable but is not satisfying. Indian-pipe is reported to have been used by Indians as an eye lotion.
Note: The plant is parasitic on roots or saprophytic on decaying vegetation and has lost all green colour.

References: Furman, T.E., 1966. Symbiotic relationships of *Monotropa. American Journal of Botany*, vol. 53, 627.

Hirce, E.G. and A.F. Finocchio, 1972. Stem and root anatomy of *Monotropa uniflora. Torrey Botany Club Bulletin.*, vol. 99(2), 89–94.

Lutz, R.W. and R.D. Sjolund, 1973. *Monotropa uniflora*: ultrastructural details of its mycorrhizal habit. *American Journal of Botany*, vol. 60(4), 339–45.

Olson, A.R., 1980. Seed morphology of *Monotropa uniflora* L. (Ericaceae). *American Journal of Botany*, vol. 67(6), 968–74.

Riley, R.K., 1971. A study of *Monotropa uniflora* mycorrhiza and host relationships. Ph.D. thesis, West Virginia University, Ann Arbor, Michigan.

Labrador tea

Latin Name: *Ledum groenlandicum*
Genus: 3 species in the northern hemisphere; 1 in Atlantic Canada
Family: Ericaceae (Heath)
Etymology: Greek *Ledon*, ancient name of a species containing aromatic resin that was thought to resemble the aroma of *Ledum*; of Greenland
Habitat: Bogs, wet barrens, wooded swamps, poorly drained pastures, and open areas
Range: Arctic America, south to Pennsylvania and Minnesota
Longevity: Perennial
Flowering: June 10–30
Features: White flowers are irregular, have 5 separate petals, and are clustered at the stem top in a showy, rounded umbel. The leathery leaves are densely rusty-woolly beneath and have strongly in-rolled edges.
Fruit: Many-seeded capsule

Height: 1 m.
Ecology: Leaves and young twigs are eaten by deer, moose, and other wildlife
Uses: The leaves can be used as an herbal tea after drying and crushing. It has a spicy evergreen flavour and possesses soothing properties. It can also be smoked in small amounts to produce a similar effect.
Note: The plant is an evergreen.
Similar Species: *Kalmia angustifolia* has pink flowers and flat leaves with a non-woolly undersurface.

References: Angelo, R., 1979. *Ledum groenlandicum* rediscovered in Concord, Massachussetts. *Rhodora*, vol.81, 285–6.

Reader, R.J., 1982. Geographic variation in the shoot productivity of bog shrubs and some environmental correlates [Leatherleaf (*Chamaedaphne calyculata*), bog laurel (*Kalmia polifolia*), Labrador tea (*Ledum groenlandicum*). *Canadian Journal of Botany*, vol.60(4), 340–8.

Rhodora

Latin Name: *Rhododendron canadense*
Genus: More than 600 species of the north temperate and Arctic zones; 2 in Atlantic Canada
Family: Ericaceae (Heath)
Etymology: Greek name, meaning rose tree; Canadian
Habitat: Common in swamps, rocky barrens, edges of bogs, poorly drained soils, and wet pastures
Range: Newfoundland to Québec, south to New Jersey and Pennsylvania
Longevity: Perennial
Flowering: May 20 – June 20
Features: The rose-purple corolla is two-lipped and funnel-shaped. The leaves are thin, deciduous, dull green, and have leaf margins nearly or entirely without teeth. This shrubby plant has wiry, wooden stems.
Fruit: Elongate capsule
Height: To 1 m.
Uses: All parts of the plant are reported poisonous.
Note: Rhodora is one of our dominant spring-flowering shrubs. It blooms before the leaves develop.
Similar Species: There are 2 forma, of which *albiforma* has white flowers and is rare.

References: Poulin, G. and C. Jankowski, 1979. Notes on the biostatic properties of the flowers of *Rhododenron canadense*. *Economic Botany*, vol.32(4), 433–4.

Sheep laurel
also called lambkill

Latin Name: Kalmia angustifolia
Genus: About 8 species, of North America; 2 in Atlantic Canada
Family: Ericaceae (Heath)
Etymology: Dedicated to Pehr Kalm, a pupil of Linnaeus; broad-leaved
Habitat: Bogs, barrens, blueberry fields, barrens, roadsides, and old pastures
Range: Labrador to Ontario, south to Georgia and Michigan
Longevity: Perennial
Flowering: June 20 – early July
Features: Plant is an erect shrub growing in clumps and has showy pink flowers in lateral clusters. Ten stamens are sunken in depressions on the 5 petals.

Fruit: Many-seeded capsule
Height: Up to 1 m.
Ecology: Although the leaves and stems are poisonous to livestock, especially sheep, deer are reported to feed on its leaves with impunity.
Similar Species: K. polifolia has terminal flowers and leaves that are in-rolled on the edge and smooth beneath.

References: Hall, I.V., L. Jackson, and C.F. Everett, 1973. The biology of Canadian weeds. *Canadian Journal of Plant Science,* vol.53(4), 865–73.
Jaynes, R.A., 1971. A gene controlling pigmentation in sheep laurel. *Journal of Heredity,* vol.62(3), 201–3.
Singh, P., 1976. Some fungi isolated from the seeds of Kalmia angustifolia (weed of forest sites, biological control). *Bi-monthly Research Notes Canadian Forest Service,* vol.32(2), 11–12.

Trailing arbutus
also called Mayflower

Latin Name: *Epigaea repens*, var. *glabrifolia*
Genus: Two species, the only other found in eastern Asia; one in Atlantic Canada
Family: Ericaceae (Heath)
Etymology: Greek *epi*, upon, and *gaea*, the earth (referring to the low growth); running on the ground; smooth-leaved
Habitat: Open lands with well-drained acidic soils such as pastures, hillsides, barrens and woodlands
Range: Labrador to Manitoba, south to Virginia, North Carolina and Tennessee
Longevity: Perennial
Flowering: April 15–May 15
Features: The waxy tubular flowers are woolly in the throat, have a spicy fragrance and occur in clusters at the ends of branches. Heart-shaped, evergreen leaves occur alternately on the woody trailing stem
Fruit: Many-seeded scarlet capsule
Height: Prostrate, with branches rising to 5 cm.
Ecology: Fruit is eaten by ants.
Uses: The raw corolla is good as a sour-sweet nibble or as an addition to salads.
Horticulture: Light, acidic soil; semi-shade
Note: This is the floral emblem of Nova Scotia and is generally recognized as the first floral sign of spring. Supposedly, this was the first flower to greet the Pilgrims when they arrived at Plymouth Rock on the *Mayflower*. The plant has often been overpicked or uprooted, so harvesting is not encouraged.

Blueberry
also called ground hurts

Latin Name: *Vaccinium angustifolium*
Genus: Possibly 150 species of the northern hemisphere; 12 species in Atlantic Canada
Family: Ericaceae (Heath)
Etymology: Latin *vaccinum*, of cows; narrow-leaved
Habitat: Poor, dry, acidic soil on exposed barrens, old pastures, railroads, dry bogs, and open woods
Range: Newfoundland to Saskatchewan, south to the uplands of Virginia and Ohio
Longevity: Perennial
Flowering: June
Features: The white- to rose-coloured, vase-shaped flowers occur in a raceme. Leaves are green on both sides, alternate, have small teeth, and may be pubescent on the midrib only.
Fruit: Blue berry
Height: Usually 10–40 cm., seldom more then 70 cm.
Ecology: The berries are used as food by many animals.

Uses: Berries are popular as a fresh fruit or baked in pies and muffins. They can also be used for jam and blueberry wine, or dried and used as a substitute for raisins and currants in many recipes.
Note: This tetraploid species is highly variable and hybridizes easily. Different clones show a wide variety of characteristics, and the plants vary considerably in different environments.
Smiliar Species: *V. Brittoni* has black berries, and the twigs and leaves are glaucous; *V. boreale* is a smaller plant; *V. myrtilloides* has densely pubescent leaves on both sides and no teeth.

References: Hall, I.V., L.E. Aalders, N.L. Nickerson, and S.P. Vander Kloet, 1979. The biological flora of Canada, I. *Vaccinium angustifolium* Ait., sweet lowbush blueberry. *The Canadian Field-Naturalist*, vol. 93, 415–30.

Vander Kloet, S.P., 1978. Systematics, distribution, and nomenclature of the polymorphic *Vaccinium angustifolium* Aiton. *Rhodora*, vol. 80, 358–76.

Fox berry
also called bog cranberry

Latin Name: *Vaccinium Vitis-Idaea*, var. *minus*
Genus: 18 species in Canada; 12 in Atlantic Canada
Family: Ericaceae (Heath)
Etymology: Latin *vaccinus*, of cows; grape of Mt. Ida; smaller
Habitat: Headlands, barrens, heaths, acid bogs, and muskegs
Range: Arctic America, south to New England; Lake Superior and British Columbia
Longevity: Perennial
Flowering: June
Features: Bell-shaped, four-lobed pink or white flowers occur in small terminal clusters. The shiny-topped leaves are alternate, obovate, with rolled edges, and have black hairs underneath.
Fruit: Red berry
Height: 10–30 cm.

Ecology: Animals, including birds, dig the berries out of the snow in winter
Uses: Berries can be eaten fresh or cooked in jam, pies, pastries, and sauces. They are high in vitamin C.
Horticulture: Moist, well-drained, acid soil (pH 4.0–5.0); either sun or shade
Note: The stem is creeping, but the branches are erect
Similar Species: Other *Vaccinium* species

References: Biermann, J.E., 1975. A description of *Vaccinium vitis-idaea*. *Fruit Varieties Journal*, vol. 29, 5–7.

Hall, I.V. and C.E. Beil, 1970. Seed germination, pollination and growth of *Vaccinium vitis-idaea* var. *minus* Lodd. *Canadian Journal of Plant Science*, vol. 50, 731–2.

Hall, I.V. and J.M. Shay, 1981. The biological flora of Canada, 3. *Vaccinium vitis-idaea* L. var. *minus* Lodd. supplementary account. *The Canadian Field-Naturalist*, vol. 95, 434–64.

Primrose

Latin Name: *Primula laurentiana*
Genus: 450 species have been described, chiefly in Central Asia; 2 in Atlantic Canada
Family: Primulaceae (Primrose)
Etymology: A diminutive of *primus*, referring to early flowering of true primrose; of the St. Lawrence Gulf and River.
Habitat: Calcareous areas such as basaltic headlands
Range: Southern Labrador, Newfoundland, and Ontario to Nova Scotia and eastern Maine

Longevity: Perennial
Flowering: Late June
Features: The white to pale lilac flowers have a five-lobed corolla and occur on a scape. Glabrous plant has basal leaves that are mealy-whitened beneath.
Fruit: Many-seeded capsule
Height: 14–50 cm.
Similar Species: *P. mistassinica* has leaves that are not mealy beneath, or are only slightly so.

Garden-loosestrife

Latin Name: *Lysimachia vulgaris*
Genus: About 140 species, mostly of central Asia; 6 in Atlantic Canada
Family: Primulaceae (Primrose)
Etymology: Named in honour of King Lysimachus of Thrace; common
Habitat: In gardens or as an escape
Range: Nova Scotia and Prince Edward Island to Ontario, south to Maryland; introduced from Europe
Longevity: Perennial
Flowering: July–September
Features: Numerous showy yellow flowers with 5 petals and in a terminal leafy panicle. The stamens are united to the top of the ovary. Plant is usually pubescent and has opposite or whorled leaves.
Fruit: Capsule
Height: 1 m. or more
Similar Species: *L. thrysiflora* occurs in wet areas and has flowers in stalks from leaf axils. *L. terrestris* is native to wet habitats and has a terminal raceme of flowers.

References: Coffey, V.J. and S.B. Jones, Jr., 1980. Biosystematics of *Lysimachua* section Seleucia (Primulaceae). *Brittonia*, vol. 32(3), 309–32.

Loosestrife

Latin Name: *Lysimachia terrestris*
Genus: 140 species, mostly of central Asia; 6 species in Atlantic Canada
Family: Primulaceae (Primrose)
Etymology: Named in honour of King Lysimachus of Thrace; terrestrial
Habitat: Boggy thickets, meadows, ditches, and marshes
Range: Newfoundland to Minnesota, south to Georgia, Kentucky, and Iowa
Longevity: Perennial
Flowering: July
Features: The yellow flowers are marked with red spots, have stamens with filaments united to the level of the ovary top, and are 1 cm. or more wide; the inflorescence is a terminal raceme.
Fruit: Capsule
Height: 0.2–1 m.
Similiar Species: *L. thrysiflora* has its flowers on dense, long-stalked oval heads in the middle-leaf axils.

References: Coffey, V.J. and S.B. Jones, Jr., 1980. Biosystematics of *Lysimachia* section Seleucia (Primulaceae). *Brittonia*, vol. 32(3), 309–22.

Starflower

Latin Name: *Trientalis borealis*
Genus: 3 species in the temperate northern hemisphere; 1 in Atlantic Canada
Family: Primulaceae (Primrose)
Etymology: Latin *triens*, one-third (referring to plant height, 1/3 foot); northern
Habitat: A pioneer in coniferous or hardwood forests
Range: Labrador to British Columbia, south to Virginia and Illinois
Longevity: Perennial
Flowering: Mid-June
Features: One to 3 white star-like flowers, with floral parts in five's, occur at the top of the stem and are subtended by a single whorl of thin pointed leaves.
Fruit: Few-seeded capsule
Height: 10–20 cm.
Note: Spreads by slender, elongate stolons and is one of the better-known woodland plants.
Similar Species: *Medeola virginiana* has parallel-veined leaves.

References: Anderson, R.C. and M.H. Beare, 1983. Breeding system and pollination ecology of *Trientalis borealis* (Primulaceae). *American Journal of Botany*, vol. 70(3), 408–15.

Sea lavender
also called marsh-rosemary

Latin Name: Limonium Nashii
Genus: About 100 species in western and eastern hemispheres; 1 in Atlantic Canada
Family: Plumbaginaceae (Leadwort)
Etymology: Greek from *leimon*, a marsh; for George Valentine Nash
Habitat: Salt marshes and around sea shores
Range: The lower St. Lawrence River and Newfoundland, south to Florida and Texas
Longevity: Perennial

Flowering: July 20 – September
Features: Small, bluish, five-petalled, papery flowers grow in a corymb on a naked scape. The thick basal leaves have petioles, dilated blades, and no teeth.
Height: Up to 30 cm.
Uses: The flower stalks are popular in dried-flower arrangements
Similar Species: Var. *trichogonum* is a more northern plant with a stiff-hairy calyx.

Swamp-milkweed

Latin Name: *Asclepias incarnata*
Genus: About 100 species in the western hemisphere; 2 in Atlantic Canada
Family: Asclepiadaceae (Milkweed)
Etymology: Greek *Aesculapius*, for whom the genus is named; flesh-coloured
Habitat: Swamps, wet thickets, and freshwater shores
Range: New Brunswick to Saskatchewan, south to Georgia and Texas, and Colorado
Longevity: Perennial
Flowering: Early August
Features: Numerous rose-purple flowers are in terminal umbels. Ten to 18 pairs of large leaves have veins curved towards the tip and are smooth or only finely pubescent underneath.

Fruit: Follicles (shown in photo) contain brown seeds with silky hairs and are 5–9 cm. long.
Height: Up to 1.5 m.
Ecology: Muskrats eat the root occasionally.
Uses: The young shoots and very young seed pods can be cooked as vegetables, or used in soups and omelettes, or braised with wild onions. Stewed milkweed pods with frogs' legs are reported to be a delicacy.
Horticulture: Moist soil; full sun
Note: The plant contains a white juice that is very irritating to the eyes.
Similar Species: *A. syriaca* has dense woolly leaves beneath.

Wild morning glory
also called hedge-bindweed

Latin Name: *Convolvulus sepium*
Genus: About 200 species, global distribution; 2 in Atlantic Canada
Family: Convolvulaceae (Bindweed)
Etymology: Latin *convolvere*, to entwine; of hedges
Habitat: Along coast, in towns, waste places, along lakeshores, roadsides, and spreading into fields and orchards
Range: Newfoundland to British Columbia, south to Florida and New Mexico
Longevity: Perennial
Flowering: Pink, five-lobed, bell-shaped flowers arise from the leaf axils. The twining stem is extensively branched and has alternate, arrow-shaped leaves.

Fruit: Four-seeded capsule
Height: 40–80 cm.
Similar Species: Forma *coloratus* is the common form. *C. arvensis* is rare and smaller.

References: Parrella, M.P. and L.T. Kok, 1979. *Oidamatophorus monodactylus* as a biocontrol agent of hedge bindweed (*Convolvulus sepium*): development of a rearing program and cost analysis. *Journal of Economic Entomology*, vol. 72(4), 590–2.

Selleck, G.W., 1979. Biological control of hedge bindweed (*Convolvulus sepium*, with the Argus tortoise beetle *Chelmorpha cassidae*) on Long Island. *Proceedings of the Annual Meeting, Northeastern Weed Science Society*, vol. 33, 114–18.

Forget-me-not

Latin Name: *Myosotis scorpioides*
Genus: More than 40 species have been identified; 5 occur in Atlantic Canada
Family: Boraginaceae (Borage)
Etymology: Greek *myos*, of a mouse, and *ous*, ear (allusion to the leaves of some species); like a scorpion
Habitat: Low, wet places such as stream banks, ditches, and meadows
Range: Newfoundland to Manitoba, south to Georgia; British Columbia; introduced from Europe
Longevity: Perennial
Flowering: Early June – July

Features: Small, blue, yellow-eyed flowers grow in a raceme and have a five-toothed corolla. The simple, untoothed leaves are alternate and have stiff, appressed hairs.
Fruit: Nutlets
Height: Up to 60 cm.
Similar Species: *M. laxa* is a smaller plant

References: Resch, J.F., D.F. Rosberger, J. Meinwald, and J.W. Appling. 1982. Biologically active pyrrolizidine alkaloids from the true forget-me-not, *Myosotis scorpioides*. *Journal of Natural Products*, vol. 45(3), 358–62.

Varopoulos, A., 1979. Breeding systems in *Myosotis scorpioides* L. (Boraginaceae). I. Self-incompatibility (in wild British population). *Heredity*, vol. 42(2), 149–57.

Sea-lungwort
also called oysterleaf

Latin Name: Mertensia maritima
Genus: About 24 species in North America; 1 in Atlantic Canada
Family: Boraginaceae (Borage)
Etymology: Named for German botanist, Franz Karl Mertens; of seashores
Habitat: Sandy beaches and gravelly shorelines above high tide level
Range: Massachusetts, north around the coast to Greenland and Alaska; Eurasia
Longevity: Perennial
Flowering: June 15 – August
Features: Blue- to rose-coloured five-petalled flowers are in a loose raceme with leafy bracts. The trailing stem has fleshy, glaucous, untoothed alternate leaves.
Fruit: 4 smooth nutlets
Height: Up to 60 cm.
Uses: The leaves are reported to have an oyster-like taste.
Note: As the flowers open one by one, the result is a delicate combination of pinks and blues.
Similar Species: Forma *albiflora*, a white-flowered form, occurs occasionally.

Skull-cap

Latin Name: *Scutellaria galericulata*
Genus: About 220 species around the globe; 2 in Atlantic Canada
Family: Lamiaceae (Mint)
Etymology: Latin *scutella*, a dish (referring to appendage on fruiting calyx); provided with little helmet-like skull caps
Habitat: Moist open locations behind sea beaches, cobbly lake borders, edges of ponds, along rivers and marshes
Range: Labrador to Alaska, south to Delaware, Ohio, and California; Eurasia
Longevity: Perennial
Flowering: July 15 – August
Features: Blue, two-lipped flowers have a conspicuous bump on the top of the calyx and occur in the axils of the leaves. The leaves are opposite, oblong, lance-shaped, wavy-margined, and have short petioles attached to the square stem.
Fruit: 4 small nutlets
Height: 10–100 cm.
Similar Species: *S. Churchilliana* has longer leaves and smaller corolla; *S. lateriflora* has flowers in an axillary, one-sided raceme.

Ground ivy

Latin Name: *Glechoma hederacea*, var. *micrantha*
Genus: 1 species in Atlantic Canada
Family: Lamiaceae (Mint)
Etymology: Greek *glehon*, for pennyroyal; ivy-like
Habitat: Around buildings, in shady places, on roadsides, or in fields
Range: Newfoundland to Alberta, south to Georgia; Pacific coast; introduced from Europe
Longevity: Perennial
Flowering: May–August
Features: The blue flowers have an irregular corolla with an upper and lower lip and are usually grouped in three's in leaf axils. Leaves are reddish, almost round, and shallowly toothed.
Fruit: 4 nutlets
Height: Creeping and trailing herbaceous stems
Uses: It was once used in France in the fermentation of beer and was popular in European countries as a medicine for aching backs.
Note: Often a bad weed around homes, where it forms large patches almost impossible to eradicate.
Similar Species: *Nepeta Cataria* has a spike-like inflorescence.

Hemp-nettle

Latin Name: *Galeopsis Tetrahit*, var. *bifida*
Genus: About 7 species of Eurasia; 2 in Atlantic Canada
Family: Laminaceae (Mint)
Etymology: Greek *gale*, a weasel, and *opsis*, appearance (refers to weasel-head shape of corolla); old generic name meaning four-parted; *bifida* is two-cleft).
Habitat: Gardens, agricultural fields, waste places, around seashores, and in towns
Range: Newfoundland to Alaska, south to North Carolina; introduced from Europe and Asia
Longevity: Annual
Flowering: July–August
Features: Flowers vary in colour from spotted purple to pinkish to white, have a two-lipped corolla, and are clustered towards the tip of the branches. The square stem is covered with bristly hairs that can penetrate human skin.
Fruit: 4 small nutlets
Height: 15–80 cm.
Note: The name, nettle, is given because of the prominent sharp-pointed calyx. This plant is a highly variable tetraploid hybrid
Similar Species: Forma *albiflora* is a white-flowered form, present especially in Cape Breton

References: Ivany, J.A., 1976. Effects of MCPA formulation and timing on hempnettle (*Galeopsis tetrahit*) control. *Canadian Journal of Plant Science*, vol.56(3), 765–7.

Hedge-nettle
also called woundwort

Latin Name: Stachys palustris
Genus: About 200 species; 2 in Atlantic Canada
Family: Lamiaceae (Mint)
Etymology: Greek *stachys*, spike; of marshes
Habitat: Wet ground, lake shores, ditches, thickets, and around seaports
Range: Newfoundland to southern Ontario, south to Nova Scotia and New England; all or mostly introduced from Europe
Longevity: Perennial
Flowering: July–August
Features: Flowers occur in axils of much-reduced leaves and have a purple two-lipped corolla. The lanceolate leaves are opposite and truncate at the base with short petioles to the square stem.
Fruit: 4 dark brown nutlets
Height: 30–50 cm.
Note: This plant is becoming an aggressive, difficult-to-control weed.
Similar Species: S. *arvensis* is an annual with ovate leaves; *Mentha arvensis* has blue flowers, while *Lycopus uniflorus* has white flowers.

Thyme

Latin Name: *Thymus Serpyllum*
Genus: More than 35 species of Eurasia; 1 in Atlantic Canada
Family: Lamiaceae (Mint)
Etymology: Greek, probably from *thyein*, to burn perfume, because of its use as incense; old generic name
Habitat: Often found in large mats along roadsides, pastures, and waste places
Range: Nova Scotia and Prince Edward Island to Ontario and North Carolina; naturalized from Europe
Longevity: Perennial
Flowering: July–August
Features: Its purple- to rose-coloured flowers have a two-lipped corolla and are crowded at ends of the branches. The plant is aromatic and has opposite small leaves and a square stem.
Fruit: Nutlets
Height: Prostrate stems to 30 cm.
Uses: It can be used as a pot-herb or, when dried and crushed, as a spice. Also, it has been reported to have medicinal qualities for relieving a cough, dispelling melancholia, and ensuring a good night's sleep.
Horticulture: Grows on well-drained, light soils
Similar Species: *Prunella vulgaris* has its flowers in an oblong, uninterrupted head.

References: van den Broucke, C.O. and J.A. Lemli, 1981. Pharmacological and chemical investigation of thyme liquid extracts (*Thymus vulgaris, Thymus serpyllum*, Labiatae, thymol; carvacrol). *Planta Medica—Journal of Medicinal Plant Research*, vol. 41(2), 129–35.

Schmidt, M., 1980. It's thyme for an adaptable herb (*Thymus vulgaris, Thymus serpyllum citriodorus*). *Organic Gardening*, vol. 27(3), 62, 64, 66.

Butter-and-eggs
also called toadflax

Latin Name: *Linaria vulgaris*
Genus: About 100 species, almost all native to Eurasia; 4 in Atlantic Canada
Family: Scrophulariaceae (Figwort)
Etymology: Latin *Linum*, the flax (referring to similarity of leaves); common
Habitat: Around towns, roadsides, railroads, and waste places, and spreading elsewhere
Range: Newfoundland to British Columbia, south to Florida and California; introduced from Eurasia
Longevity: Perennial
Flowering: July–August
Features: Flowers are 2 shades of yellow, grow in dense terminal racemes, and have a two-lipped corolla with a prominent spur at the flower base. The numerous, alternate leaves are linear to linear-lanceolate.
Fruit: Capsule
Height: 20–80 cm.
Ecology: The lips of the corolla are so firmly compressed that only long-tongued bees and certain butterflies are able to reach the nectar at the base of the spur.
Uses: It yields what at one time was considered a valuable skin lotion; its juice, mingled with milk, consitutes a fly poison.
Note: A noxious weed, spreading by rootstocks and forming dense patches
Similar Species: Forma *leucantha* has milky-white flowers except for the palate.

Turtlehead
also called balmony

Latin Name: Chelone glabra
Genus: 4 species of eastern North America; 1 in Atlantic Canada
Family: Scrophulariaceae (Figwort)
Etymology: Greek *chelone*, a tortoise (referring to flower shape); smooth
Habitat: Swamps, wet roadsides, along rocky streams and meadows, and above salt-water influence of estuarine rivers
Range: Newfoundland to Ontario, south to Georgia and Alabama
Longevity: Perennial
Flowering: July 15–August

Features: A few large white flowers occur in a dense terminal spike and are strongly two-lipped. The toothed leaves are opposite or whorled.
Fruit: Capsule
Height: Up to 150 cm.
Uses: The plant has been used for reducing inflammation, and as an anthelmintic and tonic.
Horticulture: Moist, rich, humus soil; nearly full sun
Similar Species: Var. *dilatata* has smaller top leaves, rounded to the petioles. Forma *tomentosa* leaves are densely hairy on under-side.

Common speedwell

Latin Name: *Veronica officinalis*
Genus: 250 species, mainly in Europe; 11 species in Atlantic Canada
Family: Scrophulariaceae (Figwort)
Etymology: Named for St. Veronica; of the shops
Habitat: Roadsides, cultivated fields, and shady places such as woodlands
Range: Newfoundland to Dakota, southwards; both indigenous and naturalized from Europe
Longevity: Perennial
Flowering: July–August

Features: Small lavender–blue flowers occur in racemes that grow from leaf axils and have a four-parted corolla and calyx. The plant is pubescent and has sessile, opposite leaves.
Fruit: Capsule
Height: Prostrate, creeping
Uses: It can be used as a tea substitute.
Similar Species: Var. *Tournefortii* has pointed, tipped leaves.

References: Harris, G.R. and P.H. Lovell, 1980. Adventitious root formation in *Veronica* spp. *Annals of Botany*, vol. 45(4), 459–68.

Cow-wheat

Latin Name: *Melampyrum lineare*
Genus: About 15 species, mostly in western Eurasia; 1 in Atlantic Canada
Family: Scrophulariaceae (Figwort)
Etymology: Greek *melas*, and *pyros*, wheat (referring to seed colour); linear
Habitat: Bogs, heaths, peaty or rocky barrens in rather exposed situations
Range: Newfoundland and southern Labrador to British Columbia, south to Nova Scotia, northern New England, and Wisconsin
Longevity: Annual
Flowering: July–August
Features: Pale yellow flowers often have a purple tinge, are strongly two-lipped, and occur in the leaf axils. The leaves and bracts are opposite, linear, and light green.
Fruit: Capsule
Height: Up to 20 cm.
Uses: The cow-wheat was formerly cultivated by the Dutch as food for cattle.
Note: Highly variable and arbitrarily divided into four confluent varieties in North America
Similar Species: Var. *americanum* has a bushy-branched stem.

Rib-grass
also called narrow-leaved plantain

Latin Name: Plantago lanceolata
Genus: More than 250 species, of global distribution; 6 in Atlantic Canada
Family: Plantaginaceae (Plantain)
Etymology: Latin *planta*, footprint; lance-shaped
Habitat: Hayfields
Range: In all provinces, except the Prairies, and throughout the United States; introduced from Europe
Longevity: Perennial
Flowering: June–October
Features: Small, greenish flowers occur on a naked stalk and have a membranous and minutely four-lobed corolla. The lanceolate to linear leaves are all basal.
Fruit: Capsule

Height: 30 cm.
Similar Species: P. juncoides and P. oliganthos have fleshy leaves.

References: Bassett, I.J., 1973. The plantains of Canada. *Canada Department of Agriculture Monograph*, no. 7.

Cavers, P.B., I.J. Bassett, and C.W. Crompton, 1980. The biology of Canadian weeds. 47. *Plantago lanceolata* L. *Canadian Journal of Plant Science*, vol. 60(4), 1269–82.

Primack, R.B. and J. Antonovics, 1982. Experimental ecological genetics in *Plantago*. VII. Reproductive effort in populations of *Plantago lanceolata* L. *Evolution*, vol. 36(4), 742–52.

Teramura, A.H. and B.R. Strain, 1979. Localized populational differences in the photosynthetic response to temperature and irradiance in *Plantago lanceolata*. *Canadian Journal of Botany*, vol. 57(22), 2559–63.

Sweet-scented bedstraw

Latin Name: *Galium triflorum*
Genus: There are more 300 widely distributed species, 13 of which occur in Atlantic Canada
Family: Rubiaceae (Madder)
Etymology: Greek *gala*, milk, which is curdled by some species of this genus; three-flowered
Habitat: Damp ground in mixed or deciduous woods
Range: Newfoundland to Alaska, south to Virginia and California
Longevity: Perennial
Flowering: July–August

Features: Small white flowers occur in clusters of 3 and have 4 corolla lobes but no sepals. Lanceolate leaves occur in whorls of 6 on a smooth, four-angled stem. The ovary and fruit are covered with hooked bristles.
Fruit: 2 round nutlets
Height: 20 cm. to more than 1 m.
Uses: The seeds have been used as a coffee substitute
Similiar Species: *G. asprellum* grows in a tangled, weedy mass; *G. Mullago* grows in dense clumps or colonies on roadsides and fields.

Twinflower

Latin Name: *Linnaea borealis*, var. *americana*
Genus: 2 or 3 species in the north temperate zone; 1 in Atlantic Canada
Family: Caprifoliaceae (Honeysuckle)
Etymology: Dedicated to Linnaeus, the great Swedish botanist; northern
Habitat: Cool mossy woodland, wooded swamps, and spruce bogs
Range: Labrador and Newfoundland to Alaska, south to Maryland, Indiana, and Colorado
Longevity: Perennial
Flowering: Late June
Features: A pair of delicate pink, bell-shaped flowers nod on thread-like upright stalks. The plants often form mats and have slender stems with rounded, opposite, evergreen leaves.
Fruit: One-seeded capsule
Height: Flower stalks up to 10 cm.
Ecology: The plant is sometimes eaten by white-tailed deer.
Horticulture: Partial sunlight and rich moist soil
Note: The plant was named for Linnaeus, the father of "taxonomy", with whom it was a favourite.
Similar Species: *Arctostaphylos uva-ursi* has alternate waxy leaves.

Red-berried elder

Latin Name: Sambucus pubens
Genus: About 40 species of the temperate zone and mountainous tropics; 2 in Atlantic Canada
Family: Caprifoliaceae (Honeysuckle)
Etymology: Possibly from the Greek *sambuce*, an ancient musical instrument (alluding to use of twigs for whistles or flutes); pubescent
Habitat: Rocky hillsides, along streams and edges of meadows, or as scattered plants in climax woodlands
Range: Newfoundland to Alaska, south to New England and Pennsylvania, and upland to Georgia and Colorado
Longevity: Perennial, shrub
Flowering: June 1 – June 20
Features: Numerous small white flowers borne in large pyramidal clusters. The opposite, pinnately compound leaves have 5-7 leaflets and pubescent twigs.
Fruit: Red berries
Height: 60 cm.–350 cm.
Ecology: Strong-smelling flowers attract insects.
Note: The shrub has a strong smell when bruised. The berries are reputed to be poisonous.
Similar Species: S. canadensis flowers occur in a flat-topped cluster.

References: Cram, W.H., 1982. Seed germination of elder (*Sambucus pubens*) and honeysuckle (*Lonerica tatarica*). *HortScience*, vol. 17(4), 618–19.

Joe-Pye-weed

Latin Name: Eupatorium maculatum
Genus: 500 species, with nearly global distribution; 4 occur in Atlantic Canada.
Family: Asteraceae (Aster)
Etymology: Named for Eupator, ancient king of Pontus; mottled
Habitat: Along brooks and edges of meadows and swamps
Range: Newfoundland to British Columbia, south to Pennsylvania, and in the mountains to North Carolina and Michigan
Longevity: Perennial
Flowering: Late July – September
Features: The purplish flowers have numerous florets in each head, which is grouped with others in a large flat-topped inflorescence. Leaves are attached to the purple stem in whorls of 3–6 and do not overtop the inflorescence.
Fruit: Achene
Height: Up to 200 cm.
Ecology: Seeds serve as food for some birds.
Horticulture: Moist, well-drained, and mulched soil with full sunlight
Note: According to legend, an Indian called Joe Pye used this plant to cure typhus fever.
Similar Species: Forma *Faxoni* has white flowers; var. *foliosum* has upper leaves that overtop the flowers; *E. perfoliatum* has white flowers and opposite leaves.

Goldenrod

Latin Name: *Solidago* spp.
Genus: Nearly 100 species are known, chiefly in North America; 20 in Atlantic Canada
Family: Asteraceae (Aster)
Etymology: Latin *solidus*, whole, *-ago*, probably referring to reputed vulnerary qualities
Habitat: Waste places, along roadsides, fields, and unused farmland
Range: The genus is in all provinces, but some species are restricted to certain areas
Longevity: Perennial
Flowering: Late July – October
Features: Flower heads are yellow in all species except *S. bicolor*, which is cream-coloured. Heads are few- to many-flowered with both ray and disk florets. The inflorescence varies from axillary clusters, an elongated compound inflorescence, or a flat, or round-topped corymb.
Fruit: Achene
Height: Ranges from *S. bicolor* (20–80 cm.) to *S. canadensis* (30–150 cm.)
Uses: Young leaves may be used as greens, while dried older leaves and flowers can be used for tea.
Horticulture: Well-drained, sandy, moderately acid soil (pH 5.5–6.5); full sun
Note: Closely related species tend to hybridize freely.
Most Common Species: *S. bicolor*, *S. sempervirens*, *S. canadensis*, and *S. graminifolia*

References: Melville, M.R. and J.K. Morton, 1982. A biosystematic study of the *Solidago canadensis (Compositae)* complex. I. The Ontario populations. *Canadian Journal of Botany*, vol. 60(6), 976–97.

Morse, D.H., 1982. Foraging relationships within a guild of bumble bees (*Bombus* spp., goldenrod (*Solidago* spp.). *Social Insects*, vol. 29(3), 445–54.

Pringle, J.S., 1968. The common *Solidago* species (goldenrods) of southern Ontario. *Royal Botanical Gardens (Hamilton) Technical Bulletin*, no. 3.

Blue asters

Latin Name: *Aster* spp.
Genus: 250 species, mostly in North America; at least 19 in Atlantic Canada
Family: Asteraceae (Aster)
Etymology: Greek *aster*, star (refers to flower shape)
Habitat: Variable, but most prefer open spaces such as thickets, roadsides, fields, edges of woods, and about dwellings. Some species prefer wet areas.
Range: Atlantic Canada to Ontario; some as far west as Ontario and into the United States
Longevity: Perennial
Flowering: Late July into September
Features: With 2 exceptions, Atlantic Canada's asters have blue ray florets, which are numerous on a flat or slightly convex receptacle. Leaves are simple.
Fruit: Flattened achene

Height: Ranges from *A. nemoralis* (5 cm.–90 cm.) to *A. novae-angliae* (up to 2.5 m.)
Ecology: Seeds eaten by some birds and small mammals
Horticulture: Ordinary well-drained soil and full sun, but can tolerate shade
Similar Species: *Erigeron* species

References: Brouillet, L. and J.P. Simon, 1980. Adaptation and acclimation of higher plants at the enzyme level: thermal properties of NAD malate dehydrogenase of two species of Aster (*Asteraceae*) (*Aster acuminatus, Aster nemoralis*) and their hybrid adapted to contrasting habitats. *Canadian Journal of Botany*, vol. 58(13), 1474–81.

Brouillet, L. and J.P. Simon, 1981. An ecogeographical analysis of the distribution of *Aster acuminatus* Michaux and *Aster nemoralis* Aiton. *Rhodora*, vol. 83, 521–50.

Pringle, J.S., 1967. The common *Aster* species of southern Ontario. *Royal Botanical Gardens (Hamilton) Technical Bulletin*, no. 2.

Wood aster

Latin Name: *Aster acuminatus*
Genus: 250 species, mostly in North America; 19 in Atlantic Canada
Family: Asteraceae (Aster)
Etymology: Greek *aster*, star (referring to flower's radiate head); acuminate
Habitat: Deciduous woodlands, thickets, and edges of intervales
Range: Newfoundland to eastern Ontario, south in the mountains to Georgia and Tennessee
Longevity: Perennial
Flowering: August–September
Features: The flower heads have a flat receptacle with numerous white ray florets. Plants have 10–20 leaves, with the middle or lower leaves sessile or tapering to the base.
Fruit: Achene
Height: 2–100 cm.
Similar Species: *A. umbellatus* is the only other aster that has white rays, but it is tall and occurs in more open weedy places.

References: Pitelka, L.F., D.S. Stanton, and M.O. Peckenham, 1980. Effects of light and density on resource allocation in a forest herb, *Aster acuminatus (Compositae)*. American Journal of Botany, vol. 67(6), 942–8.

Winn, A.A. and L.F. Pitelka, 1981. Some effects of density on the reproductive patterns and patch dynamics of *Aster acuminatus*. Bulletin of the Torrey Botanical Club, vol. 108(4), 438–45.

Tall white aster

Latin Name: *Aster umbellatus*
Genus: About 250 species in Eurasia, South America, and North America; 19 species in Atlantic Canada
Family: Asteraceae (Aster)
Etymology: Greek *aster*, star (referring to flower's radiate head); umbellate
Habitat: Roadsides, ditches, swamps, damp thickets, marshes, and barrens
Range: Newfoundland to Minnesota, south to the uplands of North Carolina and Kentucky
Longevity: Perennial
Flowering: Late July – September
Features: Flower heads are numerous in a flat-topped corymb with stiffly ascending branches. Both disk florets and white ray florets are present. The lanceolate to ovate leaves have no teeth, and the lower ones are sessile.
Fruit: Achene
Height: 0.2–2 m.
Horticulture: Ordinary, well-drained garden soil; full sun
Note: Wild asters are among the most abundant and beautiful of autumn wildflowers.
Similar Species: *A. acuminatus* also has white rays, but occurs in woodlands

References: Avers, C.J., 1953. Biosystematic studies in *Aster*. 2. Isolation mechanisms and some phylogenetic considerations. *Evolution*, vol. 7, 317–27.

Daisy fleabane
also called whitetop

Latin Name: Erigeron strigosus
Genus: 200 species in western and eastern hemispheres; 5 are in Atlantic Canada.
Family: Asteraceae (Aster)
Etymology: Greek *eri*, early, and *geron*, old man (an allusion to hoariness of some species); with straight, appressed hairs
Habitat: Pastures, hayfields, deserted farms, waste places, and along roadsides
Range: In all provinces and throughout most of the United States; introduced to Europe
Longevity: Annual or biennial
Flowering: Early July – September
Features: Flower head solitary to numerous, with many white ray florets and a yellow central disk. Leaves are sessile but not clasping.

Fruit: Achene
Height: 60–120 cm.
Uses: According to legend, the fleabane, when burned, was objectionable to insects. Bunches were hung in houses to keep them away.
Similar Species: Chrysanthemum Leucanthemum has a much larger flower.

References: Cronquist, A., 1947. Revision of the North American species of *Erigeron* north of Mexico. *Brittonia*, vol. 6, 121–302.

Hirose, T. and N. Kachi, 1982. Critical plant size for flowering in biennials with special reference to their distribution in a sand dune system (*Oentherra erythrosepala, Dontostemon dentatus, Erigeron strigosis,* and *Erigeron sumatrensis*). *Oecologia*, vol. 55(3), 281–4.

Pussy-toes

Latin Name: Antennaria canadensis
Genus: About 25 species in temperate North and South America; 5 in Atlantic Canada
Family: Asteraceae (Aster)
Etymology: Named from the likeness of staminate flowers to some insect antennas; Canadian
Habitat: Dry sterile soil on hills, old pastures, and deserted fields
Range: Nova Scotia to Manitoba, south to New York, and in the mountains to Virginia and Michigan
Longevity: Perennial
Flowering: May 15 – June

Features: Staminate and pistillate white florets occur in similar-appearing discoid flower heads without rays. The stem leaves are small and dominated by the basal rosette leaves.
Fruit: Achene
Height: 10–45 cm.
Ecology: Leaves and seeds are eaten by upland game birds, while deer and snowshoe hare will eat leaves.
Horticulture: Well-drained soil and full sunlight
Similar Species: A. neglecta has silky wool on upper leaf surface. See *Anaphalis*.

Black-eyed Susan

Latin Name: *Rudbeckia serotina*
Genus: 25 species in North America; 2 in Atlantic Canada
Family: Asteraceae (Aster)
Etymology: Named for Olaf Rudback, 17th-century Swedish botanist; late
Habitat: Dry meadows and roadsides
Range: Manitoba to Texas; naturalized as a weed in eastern North America
Longevity: Perennial
Flowering: Midsummer to early fall
Features: The orange flowers occur on a daisy-like head with a raised, conic-shaped, dark brown receptacle. Its lanceolate leaves are unlobed.
Fruit: Achene
Height: 1 m.
Horticulture: Sunny conditions and almost any soil
Note: It does not bloom until the second season; state flower of Maryland
Similar Species: It has been separated from *R. hirta*, but many forms and varieties have been described. Var. *serica* has crowded hairs on the under-side of the leaf; var. *lanceolata* has flowers with longer rays on its florets.

Yarrow

Latin Name: Achillea lanulosa
Genus: About 75 species in northern hemisphere; 4 in Atlantic Canada
Family: Asteraceae (Aster)
Etymology: Named for its healing powers discovered by Achilles; woolly
Habitat: Pastures, dike lands, lawns, meadows, waste places, and along roadsides
Range: Newfoundland to Alaska, south to Pennyslvania and California
Longevity: Perennial
Flowering: July 15 – September
Features: Many small greyish-white flowers occur on short peduncles in a compact corymb. The finely dissected leaves are alternate on the stem and also grow in basal rosettes.

Fruit: Achene
Height: Up to 60 cm.
Uses: It can be used in an ointment or made into milfoil tea (believed to be gifted with the power of dispelling melancholy), or the leaves can be chewed for toothache.
Similar Species: A. Millefolium is hexaploid rather than tetraploid.

References: Mulligan, G.A. and I.J. Bassett, 1959. *Achillea millefolium* complex in Canada and portions of the United States. *Canadian Journal of Botany*, vol. 37, 73-9.

Shemluck, M., 1982. Medicinal and other uses of the *Compositae* by Indians in the United States and Canada. *Journal of Ethnopharmacology*, vol. 5(3), 303-58.

Camomile
also called stinking mayweed

Latin Name: *Anthemis cotula*
Genus: Up to 60 species, native to Europe, Asia, and Africa; 3 in Atlantic Canada
Family: Asteraceae (Aster)
Etymology: The ancient name of Camomile; early united with the genus *Cotula*
Habitat: Farmyards, waste places, and along roadsides
Range: Newfoundland to southern United States; introduced from Europe
Longevity: Annual
Flowering: July–October
Features: Daisy-like flowers with yellow disk florets and sterile white ray florets. This strong-smelling plant has a branched stem and finely dissected leaves.
Fruit: Achene
Height: 10–30 cm.
Uses: Flowers are dried to make camomile tea.
Similar Species: *Chrysanthemum Leucanthemum* has larger flowers but not finely dissected leaves

---

References: Ivens, G.W., 1979. Stinking mayweed (*Anthemis cotula*, chemical control). *New Zealand Journal of Agriculture*, vol. 138(3), 21–3.

Ox-eye daisy

Latin Name: *Chrysanthemum Leucanthemum*
Genus: About 100 species, native chiefly to the northern hemisphere; 2 in Atlantic Canada
Family: Asteraceae (Aster)
Etymology: Old Greek name *Chrysanthemon*, golden flower; white flower
Habitat: Meadows, pastures, waste places, hayfields, and along roadsides
Range: In all provinces, south to New York and New Jersey; introduced from Europe
Longevity: Perennial
Flowering: June–July
Features: Flower heads solitary or few, with white ray florets and yellow disk florets. The basal leaves are coarsely and irregularly toothed, while the stem leaves are narrowly oblong.
Fruit: Achene
Height: *30–90 cm.*
Uses: Young leaves can be used in a salad
Note: When eaten by cattle, this plant gives milk a disagreeable taste, but it is difficult to eradicate from pastures.
Similar Species: *Erigeron* species have much smaller flowering heads.

References: Kumar, A., 1982. Polyploidy in *Achyranthes aspera* and *Chrysanthemum Leucanthemum*. *Acta Botanica Indica*, vol. 10(1), 151–2.

Tansy

Latin Name: *Tanacetum vulgare*
Genus: About 50 species, native to the northern hemisphere; 1 in Atlantic Canada
Family: Asteraceae (Aster)
Etymology: Of uncertain derivation; common
Habitat: Along roadsides, near old houses, and becoming a weed in fields and elsewhere
Range: Newfoundland to British Columbia, south to Georgia; introduced from Europe
Longevity: Perennial
Flowering: July–August
Features: Numerous bright yellow flower heads occur in a flat-topped corymb. The button-like head has disk florets, but no rays. The stem is smooth and has finely dissected leaves.
Fruit: Angled or ribbed achene

Height: 0.5–1 m.
Uses: Tansy was supposed to be capable of expelling intestinal worms and of stimulating the appetite. A solution was made from boiling the leaves to use in treating sprains and tired muscles. The plant has also been used as a savoury, in tansy tea, and for making tansy wine.
Note: This plant has a strong acrid odour
Similar Species: Forma *crispum* has more finely divided leaves with coarse teeth. *Senecio Jacobaea* has ray flowers.

References: Shemluck, M. 1982. Medicinal and other uses of the *Compositae* by Indians in the United States and Canada. *Journal of Ethnopharmacology*, vol. 5(3), 303–58.

Coltsfoot

Latin Name: *Tussilago Farfara*
Genus: Only 1 species, which occurs in Atlantic Canada
Family: Asteraceae (Aster)
Etymology: Latin *tussis*, a cough (this plant was a reputed remedy); Latin for coltsfoot
Habitat: Damp hillsides, riverbanks, heavy soils, and roadside cuts
Range: Newfoundland to Minnesota, south to New Jersey and Ohio; British Columbia; introduced from Europe
Longevity: Perennial
Flowering: Late April – early May
Features: Flowers appear in early spring before the leaves. The dandelion-like solitary head has both yellow disk and ray florets, which later in the season show the white bristles. Leaves are heart-shaped and woolly below.
Fruit: Achene
Height: 7–45 cm.
Uses: The leaves can be dried and burned and the ashes used as a salt substitute.
Note: Spreads by runnering root-stocks and forms large patches
Similar Species: *Petasites palmatus* has numerous silky white heads.

References: Cody, W.J. and R.B. MacLaren, 1976. Additions and rediscoveries of five plant species in Prince Edward Island. *The Canadian Field-Naturalist*, vol. 90(1), 53-4.

Ogden, J., 1974. The reproductive strategy of *Tussilago farfara* L. *Journal of Ecology*, vol. 62, 291–324.

Ragwort
also called stinking Willie

Latin Name: *Senecio Jacobea*
Genus: More than 1,000 species; 9 in Atlantic Canada
Family: Asteraceae (Aster)
Etymology: Latin *senex*, an old man (referring to hoariness of many species); of St. James
Habitat: Pastures, along roadsides, waste places, clearings, and on burnt-over ground
Range: Newfoundland to Gaspé, south to Eastern Massachusetts; Pacific coast; introduced from Europe
Longevity: Biennial or short-lived perennial
Flowering: Late July and August
Features: Numerous yellow flower heads occur in flat-topped corymb. Leaves are deeply cut into irregular segments, with basal leaves stalked and the stem leaves alternate and sessile.
Fruit: Achene
Height: 90–270 cm.
Note: Poisonous to livestock. The weed is becoming more widespread.
Similar Species: *S. vulgaris* has inconspicuous flower heads without ray florets. *Tanacetum vulgare* lacks ray florets.

Canada thistle

Latin Name: *Cirsium arvense*
Genus: About 200 species, native to the northern hemisphere; 4 in Atlantic Canada
Family: Asteraceae (Aster)
Etymology: Name *Cirsion*, used by Dioscorides; of fields
Habitat: Roadsides, fields, pastures, dike lands, and waste places
Range: Widespread throughout North America; introduced from Europe
Longevity: Perennial
Flowering: Early July – August
Features: Pinkish-purple dioecious flowers occur in numerous heads. The sessile, alternate leaves have prickly margins and are strongly decurrent.
Fruit: Oblong, flattish achene
Height: 30–90 cm.

Ecology: Caterpillars of the thistle butterfly and other closely related butterflies feed on thistles.
Uses: The leaves are edible, either raw or cooked, after the prickles have been removed and so is the stalk after the tough skin is removed. Young thistle leaves can be used in a salad.
Note: This is one of the worst weeds in the region, spreading vigorously by underground root-stocks and growing in patches.
Similar Species: Forma *albiflorum* has white flowers. *C. palustre* is less common and does not spread by its root-stocks.

References: Pedersen, L.H. and C.J. Eckenrode, 1981. Predicting cabbage maggot flights in New York using common wild plants. *New York's Food and Life Sciences Bulletin*, no. 87.

Common chicory

Latin Name: *Chichorium intybus*
Genus: About 9 species originating in Eurasia and Mediterranean; 1 species occurs in Atlantic Canada
Family: Asteraceae (Aster)
Etymology: *Chichorium*, altered from Arabic name; *Intybus*, old generic name
Habitat: Roadsides, about towns, ports, and waste places
Range: Cosmopolitan; introduced from Eurasia
Longevity: Perennial or biennial
Flowering: July–September
Features: Its large blue flowers have only ray florets, no stalks, and are in clusters of 2–3 in angles between leaf and stem. The coarse, much-branched plant has milky juice and rough, hairy basal leaves.
Fruit: Achene
Height: Up to 170 cm.
Uses: Tap-root can be used as a coffee substitute; in Europe, young roots are boiled as a vegetable. Young leaves can be used in salads or as a pot-herb and contain vitamins A and C.
Note: Chicory is slowly spreading and becoming more abundant. The flowers bloom a few at a time, last one day, turn towards the sun as the earth moves, and close on an overcast day.

References: Steiner, E., 1983. The blue sailor: weed of many uses (*Chichorium intybus*, chicory, history, medicinal, food and beverage uses, as coffee replacer or adulterant, food additive). *The Michigan Botanist*, vol. 22(2), 62–7.

Goat's beard

Latin Name: *Tragopogon pratensis*
Genus: About 50 species of Europe and North Africa; 2 in Atlantic Canada
Family: Asteraceae (Aster)
Etymology: Greek *tragos*, goat, and *pogon*, beard; of meadows
Habitat: Pastures, hayfields, waste places, along roadsides and railroads
Range: In all provinces, except Newfoundland; introduced from Europe
Longevity: Biennial to perennial
Flowering: June–July
Features: The yellow flowers are much like the dandelion. The tall, little-branched stems have erect, clasping, grass-like leaves.
Fruit: Plumed achenes
Height: 30–120 cm.
Uses: The root is pleasant and nutritious but should be picked before the flowering stem has developed. Bases of the lower leaves and young stem can be cooked as a vegetable or used in a salad.
Note: A rapidly spreading and persistent weed
Similar Species: *Sonchus arvensis* has wider, bright green leaves.

References: Hartmann, H. and A.K. Watson, 1980. Host range of *Albuga tragopogi* from common ragweed. *Canadian Journal of Plant Pathology*, vol. 2(3), 173-5.

Dandelion

also called dumbledor, faceclock, and piss-a-beds

Latin Name: Taraxacum officinale
Genus: More than 1,000 species described, many from northern regions; 2 in Atlantic Canada
Family: Asteraceae (Aster)
Etymology: From Arabic *Tharakhchakon*; of the shops, alluding to medicinal use
Habitat: Pastures, hayfields, cultivated land, lawns, waste places, and along roadsides
Range: Newfoundland to Alaska southwards; naturalized from Europe
Longevity: Perennial
Flowering: May–June
Features: Yellow flower head is solitary at the top of a slender, hollow scape and has only ray florets. The basal leaves are coarsely toothed with a prominent midrib. Plant contains sticky, milky juice.
Fruit: Elongated achene
Height: To 30 cm.
Ecology: Plant is eaten by snowshoe hare and other animals.

Uses: The whole plant is edible. Roots can be eaten as a vegetable after boiling in two waters or dried and used as a coffee substitute. Flowers can be used to make dandelion wine, while the new leaves are good in salads.
Note: One of the most common weeds in settled parts of Canada. The common name is a corruption of the French 'dent de lion'.
Similar Species: Var. *palustre* is less common and in wet areas. *T. erythrospermum* has reddish achenes.

References: Bunce, J.A., 1980. Growth and physiological characteristics of disturbance ecotypes of *Taraxacum officinale*. *Bulletin—Ecological Society of America*, vol. 61(2), 139.

Mann, H. and P.B. Cavers, 1979. The regenerative capacity of root cuttings of *Taraxacum officiale* under natural conditions. *Canadian Journal of Botany*, vol. 57(17), 1783–91.

Schmidt, M., 1979. The delightful dandelion. *Organic Gardening*, vol. 26(3), 112–17.

Solbrig, O.T., 1971. The population biology of dandelions. *American Science*, vol. 59, 686–94.

Lion's-paw

Latin Name: Prenanthes trifoliata
Genus: Up to 35 species in North America, Eurasia, and the East Indies; 3 in Atlantic Canada
Family: Asteraceae (Aster)
Etymology: Greek *prenes*, drooping, and *anthe*, flower; with three leaflets
Habitat: Along the edges of thickets or in woodlands on gravelly, sandy, acidic soil
Range: Newfoundland to Ohio, south to North Carolina and Tennessee
Longevity: Perennial
Flowering: August–September

Features: Nodding flower heads occur in a panicle and have a few bell-like florets with pale purplish-white or cream-coloured petals. The leaves are mostly petioled, with the lower ones generally three-divided or angulate.
Fruit: Cylindrical achenes
Height: 15–150 cm.
Note: Tuberous roots have an intensely bitter taste.
Similar Species: Var. *nana* is found on mossy places and around the coast and has an unbranched stem.

Paintbrushes

Latin Name: *Hieracium* spp.
Genus: Nearly 10,000 species described, mostly in Europe; 15 in Atlantic Canada
Family: Asteraceae (Aster)
Etymology: Greek *hierax*, a hawk (ancient people supposed that hawks used it to strengthen their eyesight)
Habitat: Old fields, pastures, waste places, and along roadsides
Range: Many species are widespread from Newfoundland to British Columbia and south into the United States; introduced from Europe
Longevity: Perennial
Flowering: June–Sept
Features: Flowering heads have 12 to many ray florets, mostly yellow but sometimes scarlet or orange. The plants are usually pubescent, often glandular; some species have only basal leaves, while others have scattered stem leaves.

Fruit: Smooth achene
Height: 15–75 cm.
Most Common Species: *H. aurantiacum, H. Pilosella, H. floribundum,* and *H. scabrum*

References: Raynal, D.J., 1979. Population ecology of *Hieracium florentinum (Compositae)* in a central New York limestone quarry. *Journal of Applied Ecology*, vol. 16(1), 287–98.

Thomas, A.G. and H.M. Dale, 1976. Cohabitation of three *Hieracium* species in relation to the spatial heterogeneity in an old pasture. *Canadian Journal of Botany*, vol. 54, 2517–29.

Thomson, J.D., 1978. Effects of stand composition on insect visitation in two-species mixture of *Hieracium (aurantiacum* and *Hieracium florentinum). American Midland Naturalist,* vol. 100(2), 431–40.

Vander Kloet, S.P., 1978. Biogeography of *Hieracium pilosella* L. in North America with special reference to Nova Scotia. *Proceedings Nova Scotia Institute of Science*, vol. 28(3/4), 127–34.

Glossary

achene: a one-seeded, one-celled, dry, hard fruit that does not open when ripe.

alternate: distributed, as leaves, at different positions on the stem, not opposite each other.

annual: a plant that completes its development in one year or one season, then dies.

anther: the pollen container of a stamen or male floral organ.

appressed: lying flat and close to some part of a plant, usually referring to hairs.

ascending: growing upwards or turned up.

astringent: binding, contracting.

awn: a bristle, often found on grass flowers.

axil: the angle between a leaf or a branch and the stem on which it is borne.

axillary: in an axil.

basal: at the base of the plant.

berry: a pulpy fruit with several seeds, as that of currant.

biennial: of two years' duration.

blade: the expanded, usually flattened part of a leaf.

bract: a small leaf or scale, often borne below a flower or flower cluster.

calyx: the outer floral ring, or sepals, usually green, but sometimes brightly coloured.

capsule: a dry fruit with many seeds.

caryopsis: a grain, as in the grasses.

catkin: a scaly spike of flowers of one sex.

clasping: partly or entirely surrounding a stem, used of leaf bases.

cleistogamous: descriptive of a flower that does not open and is self-pollinated.

compound: of a leaf, composed of two or more leaflets.

cordate: of a leaf, heart-shaped.

corolla: the petals or inner floral ring.

corymb: a cluster of flowers in which the pedicels arise from different points on the stem; the cluster has a flat or rounded top.

culm: the stem of a grass or a sedge.

cyme: flat-topped flower-cluster with the central flower opening earliest.

deciduous: having leaves that fall off in the autumn; not evergreen.

disk: a more or less fleshy or elevated development of the receptacle about the pistil; receptacle in the head of an Asteraceae.

dissected: divided into many segments.

drupe: pulpy or fleshy fruit containing a single seed enclosed in a hard shell or stone, as that of the plum.

drupelet: one part of a fruit composed of aggregate drupes, as in the raspberry.

elongate: stretched out, lengthened.

equitant: of leaves enfolding each other, as in iris.

escape: garden plant growing wild.

floret: a single flower, usually used of a composite head or grasses.

follicle: a fruit with a single chamber that opens along one side.

fruit: the seed-bearing product of a plant.

glabrous: without hairs.

gland: an organ that secretes sticky or resinous matter.

glaucous: covered with a bloom; bluish white or bluish grey.

globular: globe-like; spherical.

glume: a scaly bract on the floral parts of grasses and sedges.

grain: a fruit resembling an achene, but in which the seed coat and thin pericarp are fused into one body, particularly the fruit of grasses.

head: a dense cluster of flowers or fruits on a very short axis or receptacle.

herb: a plant without a woody stem above the ground.

hip: the berry-like enlarged calyx tube containing many achenes, found in the roses.

hirsute: with coarse spreading hairs.

hoary: greyish white.

indigenous: native to a country or region.

inflorescence: arrangement of flowers in a cluster.

involute: rolled inwards.

lanceolate: of a leaf, much longer than wide, broadest near the base and tapering towards the tip.

leaflet: a division of a compound leaf.

legume: a dry pod-like fruit, splitting down one or both sides at maturity.

ligule: a strap-shaped organ, as in ray florets of asters; also a collar of a grass blade.

linear: of a leaf, long and narrow with parallel margins.

lip: the main lobe of a two-lobed corolla or calyx; the odd and peculiar petal of the orchid family.

lobe: a rounded projection of a leaf or a leaf-like part of a plant.

mericarp: one of the two parts of the fruit of certain families, especially Apiaceae.

mid-vein: the central vein or rib of a leaf or other organ.

monoecious: having pistils and stamens in separate flowers on the same plant.

nectar: a sweet liquid secreted by the nectaries of plants.

node: the place on a stem where leaves grow or normally arise.

nut: a single-seeded fruit with a woody, hard outer coat.

oblong: longer than broad, with the sides nearly parallel for most of their length.

obovate: of a leaf, egg-shaped with the wide part near the tip.

opposite: borne two at a node, on opposing sides of an axis.

panicle: a branched cluster of flowers, each stalked, the lower branches longest and opening first.

parasitic: growing on and deriving nourishment from another living plant.

pedicel: the stalk of single flower in a cluster.

peduncle: stem of a solitary flower or of a flower cluster.

perennial: a plant that persists for more than two years.

perfect: of a flower, complete, having both stamens and pistil.

perianth: petals and sepals referred to together.

petal: a separate part of a corolla or inner floral ring.

petiole: stalk of a leaf.

pinnate: of a compound leaf, with leaflets arranged on each side of a common stalk.

pistil: the female part of a flower, composed of style, stigma, and ovary.

pod: a dry fruit, opening when mature.

polygamous: having both perfect and unisexual flowers on the same plant.

pubescent: covered with hairs.

raceme: a flower cluster with each flower borne on a short stalk from a common stem.

ray: in some Asteraceae, a modified marginal floret with a strap-like extension of the corolla.

receptacle: the enlarged summit of the peduncle of a head to which the flowers are attached.

recurved: curved backward.

rhizome: a modified, underground stem, usually growing horizontally.

rootstock: rhizome.

rosette: a dense cluster of leaves on a very short stem or axis.

sagittate: like an arrow head; of a leaf, with the basal lobes pointing towards the place of attachment.

saprophyte: a plant that lives on dead organic matter.

scape: a flowering stem growing from the root crown and not bearing proper leaves, as in the tulip.

seed: the ripened ovule, consisting of the embryo and its proper coats.

sepal: one of the separate parts of a calyx; usually green and leaf-like.

sessile: without a stalk.

shrub: a woody plant that remains low and produces shoots or trunks from the base.

silique: a capsule with two valves separating from a thin longitudinal partition.

simple: of a leaf, having a single blade not divided into leaflets.

spadix: a dense or fleshy spike of flowers, as in *Calla*.

spathe: a large leaf-like bract enclosing a flower cluster.

spatulate: shaped like a spatula, oblong, sometimes a little broader towards the upper end, and with a rounded apex.

spike: a flower cluster, the individual flowers of which are stalkless, borne on a common stalk.

spikelet: a secondary spike, especially in grasses and sedge.

spur: a hollow projection, usually at the base of a flower, as in the snapdragon.

stamen: the male organ of a flower.

stipule: an appendage at the base of a leaf.

stolon: basal branch that roots at nodes; often underground.

style: the part of the pistil between the stigma and ovary.

tendril: a slender outgrowth by which some plants attach themselves to objects.

terminal: at the end of a stem or branch.

umbel: a flower cluster in which all flower stalks arise from a common point.

unisexual: having male organs or female organs only.

utricle: a small, bladdery, one-seeded fruit.

valve: the units or pieces of a capsule or pod; the enlarged inner sepals in *Rumex*.

whorl: a group of three or more leaves arising from the same node.

woolly: entangled, soft hairs.

Selected References

General

Crockett, L.J., 1977. *Wildly successful plants: a handbook of North American weeds.* Macmillan, New York.

Dana, W.S., 1963. *How to know the wildflowers, a guide to the names, haunts, and habitats of our common wild flowers.* Dover Publications, Inc., New York.

Erskine, D.S., 1960. *The plants of Prince Edward Island.* Canada Dept. of Agriculture, Ottawa.

Fassett, N., 1969. *A manual of aquatic plants.* University of Wisconsin Press, Madison.

Ferguson, M. and R.M. Saunders, 1982. *Canadian wildflowers through the seasons.* Van Nostrand Reinhold, Toronto.

Fernald, M.L., ed., 1950. *Gray's manual of botany,* 8th ed. American Book Co., New York.

Frankton, C. and G.A. Mulligan, 1977. *Weeds of Canada,* Pub. #948. Canada Dept. of Agriculture, Ottawa.

Gleason, H., 1952. *The new Britton and Brown illustrated flora of the northeastern United States and adjacent Canada,* 3 vols. Macmillan Publishing Co., New York.

Luer, C.A., 1975. *The native orchids of the United States and Canada.* New York Botanical Garden, Bronx, N.Y..

Mackenzie, K., 1973. *Wildflowers of eastern Canada, Ontario, Quebec, Atlantic provinces.* Tundra Books, Montreal, and Collins Publishers, Toronto.

Montgomery, F.H., 1977. *Seeds and fruits of plants of eastern Canada and northeastern U.S.A.* University of Toronto Press.

Mulligan, G.A., 1976. *Common weeds of Canada.* McClelland and Stewart in assoc. with Canada Dept. of Agriculture, Ottawa.

Peterson, R.T. and M. McKenny, 1968. *A field guide to wildflowers.* Houghton Mifflin Co., Boston.

Roland, A.E. and E.C. Smith, 1969. *The flora of Nova Scotia.* The Nova Scotia Museum, Halifax.

Savage, C. and A. Savage, 1979. Canada's carnivorous plants. *Nature Canada,* vol. 8(1), 4–12.

Schnell, D.E., 1977. *Carnivorous plants of the United States and Canada.* John F. Blair Publishers, Winston-Salem, N.C.

Scoggan, H.J., 1978. *The flora of Canada,* 4 vol. National Museum of Natural Sciences, Ottawa.

Stack, A., 1980. *Carnivorous plants.* The Massachusetts Institute of Technology, Cambridge, Mass.

Zichmanis, Z. and J. Hodgins, 1982. *Flowers of the wild, Ontario and the Great Lakes region.* Oxford University Press, Toronto.

Uses

Berglund, B. and C.E. Bolsby, 1977. *The complete outdoorsman's guide to edible wild plants.* Pagurian Press Ltd., Toronto.

Fernald, A.L. and A.C. Kinsey, 1958. *Edible wild plants of eastern North America.* Harper and Row Publishers, New York.

Gaudet, J.F., 1979. *Medicinal and poisonous plants on Prince Edward Island.* PEI Dept. of Agriculture and Forestry, Charlottetown.

Gibbons, E., 1970. *Stalking the healthful herbs.* David Mackay Co., Inc., New York.

Gibbons, E., 1973. *Stalking the wild asparagus.* David Mackay Co., Inc., New York.

Hardin, J. and J. Arena, 1974. *Human poisoning from native and cultivated plants.* Duke University Press, Durham, N.C.

Kingsbury, J.M., 1955. *Poisonous plants of the U.S.A. and Canada.* Prentice-Hall, Inc., Englewood Cliffs, N.J.

Knap, A.H., 1975. *Wild harvest, an outdoorsman's guide to edible wild plants in North America.* Pagurian Press Ltd., Toronto.

MacLeod, H. and B. MacDonald, 1977. *Edible wild plants of Nova Scotia.* Nova Scotia Museum, Halifax.

Peterson, L., 1978. *A field guide to edible wild plants of eastern and central North America.* Houghton Mifflin Co., Boston.

Scott, P.J., 1975. *Edible fruits and herbs of Newfoundland.* Oxen Pond Botanic Park, Memorial University of Newfoundland, St. John's.

Thompson, W., 1978. *Medicines from the earth.* Gage Publishing Ltd., Toronto.

Turner, N.J., 1970. *Wild green vegetables of Canada.* National Museum of Natural Sciences, Ottawa.

Turner, N.J. and A.F. Szczawinski, 1978. *Wild coffee and tea substitutes.* National Museum of Natural Sciences, Ottawa.

Turner, N.J. and A.F. Szczawinski, 1979. *Edible wild fruits and nuts of Canada.* National Museum of Natural Sciences, Ottawa.

History

Anderson, A.W., 1966. *How we got our flowers.* Dover Publications, New York.

Bailey, L.H., 1933 (reprint 1963). *How plants get their names.* Dover Publications, New York.

Durant, M., 1977. *Who named the daisy? Who named the rose?.* Dodd, Mead; New York.

Eifert, V.L., 1965. *Tall trees and far horizons: adventures and discoveries of early botanists in America.* Dodd, Mead; New York.

Erickson-Brown, C., 1979. *Uses of plants for the past 500 years.* Breezy Creeks Press, Aurora, Ont.

Fitzgibbon, A. and C.P. Traill, 1868 (reprint 1972). *Canadian wildflowers.* Coles Publishing Co., Toronto.

Griffin, D. and I.G. MacQuarrie, 1980. The living sands, coastal dunes of P.E.I. *Nature Canada,* vol. 9(2), 42-6.

Harvey, M.J., 1973. Salt marshes of the Maritimes. *Nature Canada,* vol. 2(2), 22-6.

Haughton, C.S., 1978. *Green immigrants: the plants that transformed America.* Harcourt, Brace, Janovitch; New York.

Martin, K., 1981. *Watershed red, the life of the Dunk River, Prince Edward Island.* Ragweed Press, Charlottetown.

Myers, N., 1978. Disappearing legacy, the earth's vanishing genetic heritage. *Nature Canada,* vol. 7(4), 41–54.

Penhallow, D.P., 1887. A review of Canadian botany from the first settlement of New France to the 19th century. Part 1. *Transactions of the Royal Society of Canada.* 45–61.

Whiteside, G.B., 1965. *Soil survey of Prince Edward Island*, 2nd ed. Canada Dept. of Agriculture and PEI Dept. of Agriculture, Ottawa.

Photography

Editors of Eastman Kodak Company, 1981. *More joy of photography*. Addison-Wesley Publishing Co., U.S.A.

Kraulis, J.A., 1980. *The art of Canadian nature photography*. Hurtig Publishers, Edmonton.

Leighton, D., 1983. How to photograph wildflowers. *Nature Canada*, vol. 12(2), 26-33.

Patterson, F., 1979. *Photography and the art of seeing*. Van Nostrand Reinhold Ltd., Toronto.

Journals, Periodicals

The Canadian Field-Naturalist (quarterly). Ottawa Field-Naturalists' Club, Box 3264, Postal Station C, Ottawa, Ont. K1Y 4J5.

Canadian Journal of Botany (monthly). National Research Council, Ottawa, Ont. K1A 0R6.

Canadian Journal of Plant Science (quarterly). Canadian Society of Agronomy and the Canadian Society for Horticultural Science, 151 Slater St., Ottawa, Ont. K1P 5H4.

Ecology (bi-monthly). The Ecological Society of America, Cornell University, Ithaca, N.Y. 14853.

Nature Canada (quarterly). Canadian Nature Federation, 75 Albert St., Ottawa, Ont. K1P 6G1.

Miscellaneous

Marie-Victorin, Frère, 1964. *Flore laurentienne*. Les presses de l'Université de Montréal, Montréal.

Philips L. and R. Stuckey, 1976. *Index to plant distribution maps in North American periodicals through 1972*. G.K. Hall and Co., Boston.

Putman, D.F., 1940. The climate of the Maritime provinces, *Canadian Geographic Journal*, vol. 21, 135-47.

Saville, D., 1962. *Collection and care of botanical specimens*. Canada Dept. of Agriculture, Ottawa.

Smith, A.W., 1963. *A gardener's book of plant names*. Harper and Row, New York.

Taylor, R.L. and R.A. Ludwig, eds., 1966. *The evolution of Canada's flora*. University of Toronto Press, Toronto.

Index of Scientific Names

Achillea lanulosa 118
Acorus calamus 10
Actaea rubra 38
Ammophila breviligulata 4
Antennaria canadensis 116
Anthemis cotula 119
Aralia hispida 75
Aralia nudicaulis 76
Arisaema Stewardsonii 8
Asclepias incarnata 93
Aster acuminatus 113
Aster spp. 112
Aster umbellatus 114
Atriplex patula 28

Barbarea vulgaris 39
Cakile edentula 40
Calla palustris 9
Caltha palustris 36
Carex sp. 7
Cerastium arvense 32
Chelone glabra 103
Chichorium intybus 125
Chrysanthemum Leucanthemum 120
Cirsium arvense 124
Claytonia caroliniana 29
Clintonia borealis 12
Convallaria majalis 18

Convolvulus sepium 94
Coptis trifolia 37
Cornus canadensis 79
Cypripedium acaule 24
Daucus carota 78
Drosera rotundifolia 42
Epigaea repens, rar. *glabrifolia* 84
Epilobium angustifolium 72
Erigeron strigosus 115
Eriophorum spissum 6
Eupatorium malculatum 110
Filipendula Ulmaria 48
Fragaria virginiana 43
Galeopsis Tetrahit, var. *bifida* 99

Galium triflorum 107
Geum macrophyllum 49
Glechoma hederacea,
    var. *micrantha* 98
Habenaria lacera 25
Heracleum lanatum 77
Hieracium spp. 129
Hordeum jubatum 3
Hypericum perforatum 68
Impatiens capensis 66
Iris versicolor 23
Kalmia angustifolia 85
Lathyrus japonicus 62
Lathyrus palustris 63
Ledum groenlandicum 82
Lemna minor 11
Limonium Nashii 92
Linaria vulgaris 102
Linnaea borealis,
    var. *americana* 108
Lotus corniculatus 60
Lupinus polyphyllus 57
Lysimachia terrestris 90
Lysimachia vulgaris 89
Lythrum Salicaria 71
Maianthemum canadense 16
Malva moschata 67
Medeola virginiana 19
Melampyrum lineare 105
Mertensia maritima 96
Moneses uniflora 80

Monotropa uniflora 81
Myosotis scorpioides 95
Nymphaea odorata 33
Oenothera biennis 73
Oenothera perennis 74
Oxalis montana 64
Oxalis stricta 65
Plantago lanceolata 106
Potentilla anserina 47
Potentilla norvegica 46
Potentilla palustris 44
Potentilla recta 45
Prenanthes trifoliata 128
Primula laurentiana 88
Prunus pensylvanica 56
Ranunculus sp. 34
Rhododendron canadense 83
Rosa multiflora 54
Rosa virginiana 55
Rubus chamaemorus 50
Rubus odoratus 51
Rubus sp. 52
Rudbeckia serotina 117
Rumex Acetosella 27
Rumex crispus 26
Sambucus pubens 109
Sanguisorba canadensis 53
Sarracenia purpurea 41
Scirpus acutus 5
Scutellaria galericulata 97
Senecio Jacobea 123

Sisyrinchium monatnum,
    var. *crebrum* 22
Smilacina racemosa 13
Smilacina stellata 15
Smilacina trifolia 14
Solidago spp. 111
Spergularia rubra 30
Stachys palustris 100
Stellaria graminea 31
Streptopus roseus,
    var. *perspectus* 17
Tanacetum vulgare 121
Taraxacum officinale 127
Thalictrum polygamum 35
Thymus serpyllum 101
Tragopogon pratensis 126
Trientalis borealis 91
Trifolium pratense 58
Trifolium procumbens 59
Trillium cernuum 20
Trillium undulatum 21
Trilochin elata 2
Tussilago Farfara 122
Typha latifolia 1
Vaccinium angustifolium 86
Vaccinium vitis-idaea,
    var. *minus* 87
Veronica officinalis 104
Vicia Cracca 61
Viola cucullata 69
Viola pallens 70

Index of Common Names

American beach grass,
    see marram grass
Arrow-grass 2
Avens 49
Bakeapple 50
Balmony, see turtlehead
Beach pea 62
Bird's-foot-trefoil 60
Bird cherry, see pin cherry
Blackberry 52
Black-eyed Susan 117
Blue asters 112
Blue flag 23
Blue violet 69
Blue-bead lily 12
Blue-eyed grass 22
Blueberry 86
Bog cranberry, see fox berry
Bristly sarsaparilla 75
Broad-leaved cat-tail 1
Bunchberry 79

Butter-and-eggs 102
Buttercup 34
Calamus, see sweet flag
Camomile 119
Canada thistle 124
Canadian burnet 53
Clintonia, see blue-bead lily
Cloudberry, see bakeapple
Coltsfoot 122
Common chicory 125
Common Lady's slipper,
    see stemless Lady's slipper
Common St. John's wort 68
Common speedwell 104
Common wild rose 55
Corn-lily, see blue-bead lily
Corpse-plant, see Indian-pipe
Cottongrass, see hare's-tail
Cow parsnip 77
Cow-wheat 105
Cracker berry, see bunchberry

Crowfoot, see buttercup
Curled dock 26
Daisy fleabane 115
Dandelion 127
Dumbledor, see dandelion
Evening primrose 73
Faceclock, see dandelion
False Solomn's seal 13
Field chickweed 32
Fireweed 72
Flowering raspberry 51
Forget-me-not 95
Fox berry 87
Foxtail barley 3
Garden-loosestrife 89
Garden-lupin 57
Giant bulrush 5
Goat's beard 126
Goldenrod 111
Goldthread 37
Ground hurts, see blueberry

Ground ivy 98
Hare's-tail 6
Hedge-bindweed,
    see wild morning glory
Hedge-nettle 100
Hemp-nettle 99
Indian cucumber root 19
Indian-pipe 81
Indian turnip,
    see Jack-in-the-pulpit
Jack-in-the-pulpit 8
Joe-Pye-weed 110
Labrador tea 82
Lambkill, see sheep laurel
Large willow-herb,
    see fireweed
Lesser duckweed 11
Lily of the valley 18
Lion's-paw 128
Liverberry,
    see rose twisted-stalk
Loosestrife 90
Low hop-clover 59
Marram grass 4
Marsh cinquefoil 44
Marsh marigold 36
Marsh-rosemary,
    see sea lavender
Meadow rue 35
Moccasin flower,
    see stemless Lady's slipper
Musk-mallow 67
Narrow-leaved plantain,
    see rib-grass
Nodding trillium 20
One-flowered shinleaf 80
Orach 28
Ox-eye daisy 120
Oysterleaf, see sea-lungwort
Paintbrushes 129
Painted trillium 21

Pigeon berry, see bunchberry
Pin cherry 56
Piss-a-beds, see dandelion
Pitcher-plant 41
Primrose 88
Purple loosestrife 71
Pussy-toes 116
Queen Anne's lace 78
Queen of the meadows 48
Ragged fringed orchid 25
Ragwort 123
Red baneberry 38
Red clover 58
Red-berried elder 109
Rhodora 83
Rib-grass 106
Rose twisted-stalk 17
Round-leaved sundew 42
Rough cinquefoil 46
Rough-fruited cinquefoil 45
Sand reed, see marram grass
Sand-spurry 30
Sea lavender 92
Sea-lungwort 96
Sea-rocket 40
Sedge 7
Sheep laurel 85
Sheep-sorrel 27
Silverweed 47
Skull-cap 97
Small white violet 70
Spotted jewelweed,
    see spotted-touch-me-not
Spotted-touch-me-not 66
Spring-beauty 29
Squirrel-tail grass,
    see foxtail barley
Starflower 91
Starry false Solomon's seal 15
Stemless Lady's slipper 24
Stinking mayweed, see camomile

Stinking Willie, see ragwort
Stitchwort 31
Sulphur cinquefoil,
    see rough-fruited cinquefoil
Sundrops 74
Swamp-milkweed 93
Sweet flag 10
Sweet-scented bedstraw 107
Tall white aster 114
Tansy 121
Thimbleberry,
    see flowering raspberry
Three-leaved false Solomon's
    seal 14
Thyme 101
Toadflax, see butter-and-eggs
Trailing arbutus 84
Tufted vetch 61
Turtlehead 103
Twinflower 108
Water arum 9
Water-lily 33
White rose 54
Whitetop, see daisy fleabane
Wild carrot,
    see Queen Anne's lace
Wild lily of the valley 16
Wild morning glory 94
Wild pea 63
Wild sarsaparilla 76
Wild strawberry 43
Winter cress, see yellow rocket
Wood aster 113
Wood sorrel 64
Woundwort, see hedge-nettle
Yarrow 118
Yellow wood sorrel 65
Yellow rocket 39